# Water Stewardship

# Water Stewardship

Pernille Ingildsen

**Published by**          **IWA Publishing**
**Alliance House**
**12 Caxton Street**
**London SW1H 0QS, UK**
Telephone: +44 (0)20 7654 5500
Fax: +44 (0)20 7654 5555
Email: publications@iwap.co.uk
Web: www.iwapublishing.com

First published 2020
© 2020 IWA Publishing



*British Library Cataloguing in Publication Data*
A CIP catalogue record for this book is available from the British Library

ISBN: 9781789060324 (paperback)
ISBN: 9781789060331 (eBook)

Cover image by Neil Rosenstech
Cover design by Haline Ly

# Contents

*To my helpers on this quest Tina Monberg and
Gustaf Olsson who helped me navigate and with
whom this work is entirely entangled*

# About the Author

Pernille Ingildsen co-authored *Smart Water Utilities – Complexity made Simple* with Professor Gustaf Olsson in 2016 as well as 'Get More Out of your Wastewater Treatment Plant' in 2001. She holds a PhD from Lund University, where she wrote her thesis 'Realizing Full-Scale Control in Wastewater Treatment Systems Using In Situ Nutrient Sensors'.

Pernille has throughout her career been dedicated to bridging theory with practice to obtain water sustainability. She has been working in academia, utilities, consulting companies and product companies. Throughout she has kept close contact with academia and continually contributes with input from real-life applications as well as taking ideas and input from academia and applying them in practice.

She was recently appointed co-Editor in Chief for IWA Publishing journal AQUA. She has been a key-note speaker at Leading Edge Technology conference in 2016 and at IWA New Development in IT and Water Conference in 2016 as well as a panellist at the World Water-Tech Innovation Summit in 2017 and the WEX Global 2018 conference.

Pernille works at Kalundborg Utility in Denmark, the home of the world's largest and oldest industrial symbiosis. Here she led major innovative projects. Most notably by establishing the water treatment plant Tisso II, that treats surface water to drinking water standards without the use of chlorine.

# Preface – A Guide to the Reader

This book is for everybody who shares a dream of moving from water professionalism to water stewardship. The book attempts to answer what that would mean and how that could come about and why it is important.

The content of this book is divided into ten chapters following the proportions of Fibonacci numbers, refound everywhere in nature. Overlaying the proportions on a shell as below illustrates how the book attempts to make a journey into the centre of the issue of water stewardship, taking steps forward and becoming briefer and more condensed on the way.

This means that Chapter 1 is the longest in this book and takes up a lot of space. It has been a difficult chapter to write, and it is also the most difficult chapter to read – my reviewers have told me. The chapter attempts to capture something that is 'invisibly in plain view'. It describes a pervasive sense of 'something is off'. Something that is so difficult to capture and describe because it is ingrained in the very way we think and interact with each other as well as the way we interact with the nature around us and within us. As in the movie The Matrix, it is like Neo's sense of 'something is wrong with this world'. It has taken time to decipher it, and it takes time pointing it out so that I understand it. And I hope this pointing out will resonate with something in you.

Chapter 2 provides a practical antidote to the first chapter's difficulties. While one may wonder in the first chapter if these senses of worries lead anywhere, Chapter 2 explains how the emerging insights lead to practical changes in

© IWA Publishing 2020. Water Stewardship
Author: Pernille Ingildsen
doi: 10.2166/9781789060331_xi

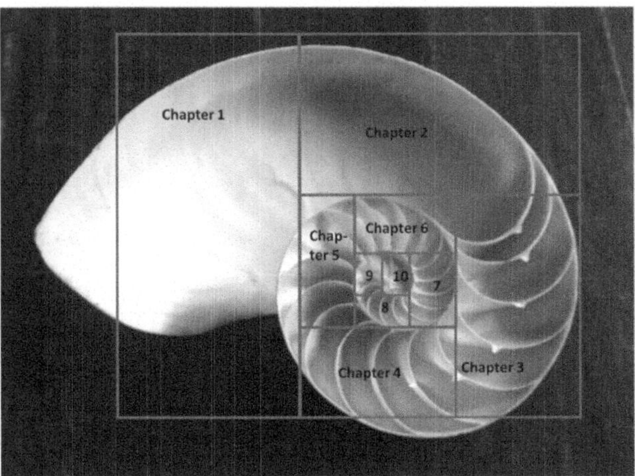


**Figure 1** Illustration of the principle of Fibonacci numbers in nature (based on photo by Roan Lavery, Unsplash).

approach to the work with water, how it changes the way projects are led, and how water challenges are approached. It also explains how the practical work with water is a two-way street where the work informs the philosophical perspective and deepens understanding of the complexity of integration and a more respectful approach to water.

This developmental process is compared to the theoretical framework of spiral dynamics excavated in our changes of mindsets by Dr Clare W. Graves in Chapter 3; a model that explains how we have seen the world differently as we and our societies have developed through a human evolutionary process of thought and sense. His theory explains how we at this stage are standing in front of a major leap forward as we progress from first-tier stages to a new set of second-tier stages. A transition that makes sense in the struggle described in Chapters 1 and 2 to see the world of water differently. It becomes clear that the urge to change fundamentally is not a water urge but a global human urge to develop.

Looking back into the recent international development of concept like integrated water resource management (IWRM), the patterns identified by Graves in Chapter 3 are discernible in the development of our leading-edge understanding of how we must work with water. A number of approaches are described including the Sustainable Development Goals, the idea of water as a common good, water footprint, IWRM and the work by the Alliance for Water Stewardship in Chapter 4.

As it has become apparent that new models are required to think differently, to enable a kind of 'thinking with the future', such tools are presented in Chapter 5.

Models for collective collaboration as well as new models for personal practices are presented.

As we approach the latter chapters of the book, they become gradually shorter and shorter. Chapter 6 tries to elicit some blind spots and increases focus on the concept of blind spots; the ideas we take for granted and do not even identify as an entity in ourselves or our culture. As these invisible concepts appear in our thoughts as separate ideas rather than as part of our operating system, they open up new possibilities – and new worries. They open a space for possible change. I have seen only a few; please keep searching.

Chapter 7 describes an interesting example of a utopian vision, written a hundred years ago by Charlotte Perkins Gilman. She envisioned a sustainable society that was characterised by both development of human nature and the nature in which the society existed. A key feature was the focus on future generations, the beautiful upbringing of children being the core value. I like this vision because it does not have a sense of austerity or strong morality for succeeding, rather it has love in all dimensions; not 'unicorn and rainbow' love or romantic love, but real fundamental love as it can exist with gratitude, truth, compassion, intelligence and humbleness.

However, perhaps we can all work on and contribute to the sustainability vision. In Chapter 8, Donella Meadow in her potent speech 'Down to earth' teaches us important lessons about our innate abilities to visioning.

Chapters 9 and 10 are at the centre of the Fibonacci shell as illustrated above. Here, it is all wrapped up. I leave it to you to find out how it all ends.

A reviewer asked if this is a 'self-help book for water professionals'. Perhaps it is kind of that. A book for reflecting on our role with water at this pivotal time in water history. Throughout the chapters I have included 'questions for reflection'. These are meant as small breaks for reflection. I hope they can work as such. If they don't work for you, skip them. The book contains a lot of literary quotes of authors who can express themselves much better than I. I hope these quotes will support the comprehension of the emotional fabric of the thinking.

In my work with this book Professor Gustaf Olsson and Mediator Tina Monberg have been my invaluable wise men and women. They have helped me merge the technical field with what I could call 'the human aspirational field'. I hope it becomes apparent what I mean as you read through the book.

I hope this will all be of value to you.

# Forewords

---

## FOREWORD BY GUSTAF OLSSON

During more than two decades I have followed Pernille's career as a water professional. Her PhD thesis almost 20 years ago is still cited; she has developed innovative full-scale operations of treatment plants; she has published her work in prestigious journals and been invited to talk at international conferences. She has published a widely praised book on 'complexity made simple' and together we put a lot of work to develop the book on how treat water 'smarter'. She is responsible for strategic planning within a progressive utility. Most people would be more than comfortable to have achieved all of this. Still, over the years Pernille and I have talked about that 'something is missing'. As she expresses:

> *'Exhausting myself in attempts to make things better, to make things work, to make the feeling go away. But despite the enormous energy I put into my work, I lacked the inner sense of happiness with the results. They did not ring clear-true with me.*
>
> *I felt that while I had an understanding of how "to do water smart", I missed a different dimension. It had been missing all the while, but it was first now that I could discern it. My troubling understanding was that I lacked the ability "to do water" in harmony with my emotional landscape and my more "spiritual" aspirations.'*

This book is different from any other text written by a water professional that you have read. It is a game-changer and because of that it takes an effort to comprehend and fully appreciate the immense implications of the book's message. It requires your reflection and it will change your thinking. As water professionals we mostly consider water as a commodity, characterized by concentrations of a multitude of components. The raw drinking water we extract from lakes, rivers, or aquifers should satisfy a certain quality, but how do we look at the water source itself? We deliver, hopefully, a sufficiently clean effluent from treatment plants to the receiving waters. Do we consider all possible aspects of the health of the receiving water as we return our used water? And what is our reaction and our attitude to the impact of all poorly treated water?

This new thinking does not always create a comfortable feeling, when we realize that our industrialisation and wasteful lifestyle will bring about a lot of misery. However, this can also lead us to another dimension of our actions. Wholeheartedly gets a new meaning. It is a matter of justice, of responsibility to our children and grandchildren, and to nature.

As we realise that the water issue is becoming the most urgent challenge of this century we are forced to think differently. We have created a lot of problems and now we must find other ways of thinking to solve these problems. It is no longer enough to consider water as a substance defined by volume, flow rate, concentrations and compositions. Water is so much more, and it strikes our heart. It is not only that our life depends on it, but our senses and heart are deeply affected by water. Pernille is at the forefront of water research and technology and at the same time describes that water is so much more than a commodity.

Many cultures have been developing around water as the central source of life. In our industrialised society we seem to have forgotten this and simply take water availability for granted. Still we have got quite a few reminders, from droughts, from polluted waters, from misuse of water. We are not protected from water scarcity.

In my book *Water and Energy* I told of an experience many years ago that has influenced my thinking since then. My wife and I visited Morocco for a short holiday during the winter season. Instead of staying on the beach we went to discover more of the interesting country. We had heard about the Blue People from the Sahel region, so we went to Guelmim (Gulimin), found a local guide at the street and continued to an oasis at the edge of the Sahara Desert. The Blue People had arrived on camels from far away and stayed in the oasis to trade goods they needed. A little pond in the oasis, around ten meters across, provided the difference between life and death. We were invited into the tent of a proud representative of the people. We had tea together and talked via our interpreter for hours. One of the first questions from our host was: 'do you have water at home?' 'Yes', we replied. 'Do you have sufficient water for your cattle?' he continued. Thinking about the abundance of water in Sweden and the clean beautiful lake close to our little summer cottage, I did not know how to answer the question properly, but said: 'yes, the cattle have enough water'. I did not dare to mention that our lake has drinkable water. The immediate reply was: 'why then did you come here?' The question has followed me since then: how would you properly answer his question and not feel guilty? Having clean water available is a sign of extravagant wealth. Too often we take it for granted.

One of the sustainable development goals (SDG) is SDG6: clean water and sanitation. Clean accessible water for all is an essential part of the world we want to live in. Water scarcity affects more than 40% of the world's population. It is apparent that if enough clean water is not available it will have an impact on poverty, health, food production and many more SDGs.

In our culture of consumerism, we have forgotten to treat water with respect. In many countries people waste huge volumes of drinking water while people living in other regions do not get enough water to drink. Not only humans but animals and plants are dying due to lack of clean water. We know that for a thirsty man a drop of water is worth more than a sack of gold. Nature should not be valued according to the instrumental value for humans but has its own right to exist. If we allow ourselves to reflect about nature around us, we will care for it with great respect. All places are sacred until treated with disrespect. So, as Pernille says: 'if we genuinely believe that water is sacred, this would be a world of grief.'

A fundamental attitude should be that we humans can never control nature. We are guests, we are invited to enjoy gifts from nature. How do we do this in a responsible way? Do we truly understand the meaning of sustainability, not to exhaust the gifts of nature but to leave the Earth, our home, undestroyed to our children? As Barbara Ward expresses it: 'we have forgotten to be good guests, how to walk lightly on the earth as its other creatures do'. Many problems in the world are caused by the difference between how people think and act and how nature works.

A few years ago, I had the privilege to meet and discuss with the legendary Chief Oren Lyons, faith keeper of the Wolf Clan, Onondaga Council of Chiefs in North America. He reminded me about the 'seven generation decisions'. How do we live responsibly and respectfully? Do our decisions take our future generations into consideration?

In this book you will follow Pernille's personal struggle to find a way that is unknown, outside any professional training and still so essential for a professional and meaningful life. She demonstrates how we can integrate heart and head in new rewarding ways in a technical culture where science and objectivity are the golden standards. It is a step further towards the original source of motivation and meaning. As Pernille so clearly tells us: we should not cease to dream.

Read the book carefully. It will take a great effort to fully appreciate the significance of it, but when you realize the importance of the message in the book your outlook will be radically changed. You will never more look at water as a commodity, but a precious gift that your life depends on, every day.

*Vraangoe, in the Gothenburg archipelago, Sweden, January 2020.*
**Gustaf Olsson**

## FOREWORD BY TINA MONBERG

*'A dreamer is one who can only find his way by moonlight, and his punishment is that he sees the dawn before the rest of the world.'*
Oscar Wilde

This book offers you a new playful and loving relationship to water far away from our present narrative. It gives us a gentle wakeup call by showing us a wise way forward without banging us on the head with shame and doomsday prophecies.

We live in a competitive society where faster, better or cheaper are the main buzzwords, that drives our agenda. Many of us, being in the hamster wheel, have lost our capacity to see the difference between what we are doing, and what is natural. In many ways, we have gone to sleep by not seeing what is and therefore, a kind of ignorance is driving us. Or how could we see it differently, when we see the consequences of our way of navigating on planet earth. Our present relation to water is a competitive story of the use of water. The other day I heard a businessman going further by using Dante's seven sins to describe our time and ways we relate to water: lust, wasting, greed, laziness, anger, jealousy or vanity.

The question that I continue to ask myself is 'living in 2020, couldn't we do better?' In 1969 the futurist Buckminster Fuller asked for an 'Operating Manual for Spaceship Earth' in his book by the same title. Fuller describes Earth as a spaceship flying through space with a finite amount of resources that cannot be resupplied. If we don't find a way to an infinite relationship to earth and her water, we have no future. The basic of life is water, and therefore we desperately search for other planets holding water to have an alternative to live on this planet.

The book you are holding in your hand is a new manual for our spaceship Earth to understand water and take care of the water we already have. It is daring, going beyond the mainstream narrative for water, caring for this sacred resource as the first of our four classical elements, and sharing, describing the situation as it is without sugar-coating it – in this book you get healthy nutrition not only for your head but also for your heart.

Pernille's story is a story of collaboration with water, as the late mathematical genius David Bohm said in the presentation 'From Fragmentation to Wholeness'; collaboration is our natural mode of operating, competition is a mistake. The story of competition and the story of collaboration are two totally different stories, as this book will show.

If we continue to use the competition story, it will lead us in a direction we don't want to go, and the competition way of thinking will lead us to the same kind of result and the same kind of knowledge. We desperately need this new manual, which has a totally different starting point. Namely, connecting to the in-depth wisdom that is not in plain sight, but when you listen deeper, you can hear it, or you can lend the ears of this book.

To me, the difference between knowledge and wisdom is that wisdom is eternal and shows us a way of being in life – while knowledge is finite, being changed all the time. Knowledge has to be rewritten all the time – while wisdom is our sounding board for securing that our decision is based on a natural ground. Knowledge without wisdom is like water in the sand.

Going beyond the famous Einstein quote; that you cannot change anything by using the same way of thinking, will demand of us to include our eternal wisdom coming from the heart. Today we know that more information is shared by the heart to the brain than opposite. So how can you reach wise solutions without using your heart? This is the question that Pernille raises in her book and gives you an answer to, showing us a way to Plato's calling for the triad; the good, the true and the beautiful.

Use Pernille's wise words to describe our, until now, indescribable relationship to water. Words and concepts that can facilitate us going from our internal personal monologues to external shared dialogues and, thereby, we will find our new manual to spaceship earth, and our collaborative water story, as the late professor and myth specialist Joseph Campbell called for in the 'Power of Myth':

'… *The only myth that is going to be worth thinking about in the immediate future is one that is talking about the planet, not the city, not these people, but the planet, and everybody on it.'*

And what it will have to deal with will be exactly what all myths have dealt with – the maturation of the individual, from dependency through adulthood, through maturity, and then to the exit; and then how to relate to this society and how to relate this society to the world of nature and the cosmos…

And this would be the philosophy for the planet, not for this group, that group, or the other group. When you see the earth from the moon, you don't see any divisions there of nations or states. This might be the symbol, really, for the new mythology to come.'

*Gilleleje, Denmark, December 2019*
**Tina Monberg**

# Acknowledgements

Professor Gustaf Olsson and Mediator Tina Monberg have been my steady companions throughout these three years of writing this book. Whenever I doubted the project, they were there ensuring me that it would all make sense in the end. They have been important sparring partners – throwing ideas forth and back. And with their very different backgrounds they have been important 'sounding boards' for the ideas presented in the book. During the last stages they have commented on the actual text and with their ideas given the book an additional upgrade that I am very grateful for. Thanks for all the support, Gustaf and Tina.

Thanks also to the CEO at Kalundborg Utility Hans-Martin Friis Moller for being supportive during all the actual water stewardship experiments described in Chapter 2: the innovative ideation phases, the difficult implementation phases and the happy success parts as well. Hans-Martin has been there all the way and ensured that the ceiling was high up. Thanks also to every single one of my colleagues and every stakeholder, contractor and consultant, for joining me on the journeys, for enriching the ideas of water stewardship and most importantly making them real. It has been great working with all of you – both uphill and downhill.

Thanks also to Michael Stubberup and Steen Hildebrandt and all my co-students at the Sustainable Co-Creation course. Thanks for inspirational discussions about everything that sustainability entails. A special thanks for facilitating the invention of a new kind of language and expression about sustainability.

The global water community is special. I think regardless of where in the world I go, there will be water people ready to discuss water issues and water technology. It is such a generous sector to be part of, where friends are found everywhere. A special thanks to the Danish water communities, not least Danva for all the meeting spaces. I am also very grateful for all my co-nerds at the ICA community. Thanks also to everyone at IWA Publishing for all the support for the books and the editorship.

Thanks also to all the poets, the scientists, the free thinkers, the activists, the intellectuals and the dreamers whose ideas and texts this book rests upon. All those who inspired me both those who I have quoted in this book and those who I did not find space for in this book, but who are in my heart and have coloured this whole mindset of water stewardship.

Thanks to my Australian friend Haline Ly for designing a great front cover that illustrates the essence. To imagine that we should do this work together at the same time as large parts of Australia are on fire in one of the greatest sustainability crises the world has yet witnessed would be unimaginable if it was not for the fact that it is actually happening.

Thanks to my mother, the librarian, who kindled my interest in literature and writing – with whom I have spent uncounted days in the library. Thanks to my father, the sailor, who kindled my heart for water and the ocean – with whom I have spent uncounted days sailing.

# Prologue

*The thought 'We are doing it wrong' kept sneaking in on me at the most inconvenient times and places. The mounting pressure of this thought was like a scratching wool shirt on me. I suspected the good I intended to do was not as good as I had hoped. Could even be harming.*

*Such thoughts are like imagined ghosts. A fast glance under the bed always assured me that there was nothing there. I reassured myself I was doing my best; quick to leave the troubling feelings behind. And I continued into the noise and rumble of everyday professional life. I felt I was pushing boundaries in all directions. Exhausting myself in attempts to make things better, to make things work, to make the feeling go away. But despite the enormous energy I put into my work, I lacked the inner sense of happiness with the results. They did not ring clear-true with me.*

*As the ghosts kept reappearing and I kept de-imagining them, a weariness came over me. It felt as if a life force inside me was dwindling. Hardly noticeable from day to day but accumulating. To a point where I realised that I had to confront the problems head on.*

*So, in this safe space of letters on pages, I now start a new attempt in my search for a better relationship with water. An attempt of a different nature. I will muster the courage to call upon the ghosts. Attempt to break off from the main road and push my way into the roadside wilderness. Try to find a new pathway where I have not walked before – in the search for 'it'.*

*December 2016*

I had just finalised the book *Smart Water Utilities* with Professor Gustaf Olsson when I reached the above conclusion. We had found a simple way to integrate sensors, models, controls, automation, goal hierarchies etc. into an easy

© IWA Publishing 2020. Water Stewardship
Author: Pernille Ingildsen
doi: 10.2166/9781789060331_xxiii

conceptual model. The conceptual model of Measure-Analyse-Decide effectively organised the 'smart' in 'smart water utilities'. We interpreted the 'smart'-component as a brain-like layer on top of the physical assets of water and wastewater utilities.

This structure made it clear how to integrate the water cycle; seamlessly integrating the natural and the societal part of it. The concept organised the change towards 'smart water utilities' into a step-by-step process. I was pleased with the result because it had that satisfying feeling of something falling into place. Everything had come together and found rest in a larger satisfying pattern. Years of working with water in different contexts and from different perspectives finally found resolution and rest.

I think that the calmness that followed the publication of the book allowed for this new different feeling to surface. I felt that while I had an understanding of how 'to do water smart', I missed a different dimension. It had been missing all along, but it was not until now that I could perceive its absence. My troubling understanding was that I lacked the ability 'to do water' in harmony with my emotional landscape and my more 'spiritual' aspirations. There was an unintended 'coldness' to the rational intelligence of 'smart water utilities'. I had this unspoken feeling of something out of place, something lacking, something not seen.

When I opened this question to Gustaf Olsson, he responded:

*'Water is something holy. We may say that there are many things that our life depends on. Water, food, comfort, some energy … Still, water is so special. Look at the sea surface, and you will get amazed. It does not have to do with the extension of the area – you will never feel the same if you see a huge dry field. The scale of water is not important. Think about the very incomparable sacred feeling that you will get in the Scandinavian mountains when you can drink the water from a little creek. If there is water close by, you are attracted to go there or sit there, even if it is a very small stream. And, as you know, the Bible talks about "living water", and everybody at that time understood the meaning of it.*

*I have often wondered: does it make any difference even if I would find the best control method in the world for treating wastewater? Well, if my purpose is to experience to recover that "holy" water what should be my path? In professional life, we often look at water more like a commodity than the matter that all life is depending on. Of course, it is important to treat it well – still, that water is not as "holy" as the clear water you experience in the little creek.*

*Once many years ago, I learnt from a Japanese monk how the water in one little bucket was used wisely to the last drop in his morning ceremony. First to drink, then to wash the body piece by piece, all the time using the same water but for uses that asked for less and less quality of the water. Every drop was used in the best possible way. We have so much to learn!'*

This was not a topic we had explored before, so I was happy to hear that my 'problem' also resonated with my mentor for many years.

However, I would probably not have taken my discomforts seriously to the extent of writing a book, if not for Tina Monberg. Tina Monberg is a visionary mediator and has followed me on this path for several years now. She has the rare ability to create a safe space, where doubts and discomforts can be explored. At the time, Tina was helping me bring mediation into my leadership role. Mediation is a process of peaceful conflict resolution based on natural decision processes. I wanted to integrate this mindset into my leadership work as a basic principle for the resolution of conflicts on all layers: personal, interpersonal and systemic.

Kalundborg Utility was going through a significant change process; a process that posed complex questions of systemic as well as personal and interpersonal character. I was determined to succeed with my leadership role in a mild, generative and peaceful way. The process of learning mediation was challenging. It required me to understand my internal workings as well as gaining a deeper level of empathy for other's feelings. With this work, I was 'swimming upstream in myself'. I worked backwards and inwards to understand my drives most importantly in relation to my water vocation. When I finally came around to formulating this, she responded:

> '*I sense that more and more people are looking to the horizon, looking for a glint, a wisdom, an insight, and I believe that this search for the subtle in the extraordinary world will enable us to effect a collective emergence, so that we as a community will be able – on the spur of the moment – to see and manifest a new image, that we can share with each other.*
>
> *Until then, we must be compassionate with ourselves as we bang our heads walking into invisible walls.*'

The responses from Gustaf and Tina strengthened my resolve to delve into this question, to find out how to move into a better relationship with water; a relationship that is more graceful, integrated and whole. A relationship where we show gratitude, respect and modesty. Throughout the search for this new relationship to water, Gustaf and Tina have been supportive companions. In a sense, they are invisible co-authors to this book.

Through the process, I came to understand that one thing is doing things 'smart', but even if intelligence in the form of smart, big-data, industry 4.0 etc. is a huge step forward, this in itself will not be enough. There is a higher meaning with our work life, a higher purpose and a moral obligation. At this point in history, we must do better than 'smart' or 'intelligent'. We need to hit a higher tune of poetic beauty in what we do.

Money has been an effective organising principle for collaboration across the globe. Science has provided a tremendous learning experience that has propelled us into a new more profound understanding of the world, unprecedented in our evolutionary history. 'Unprecedented' seems to be a modest word for the revolution that has happened within the last few generations; a revolution that has left no aspect of human life untouched. And to our horror has left no aspect of natural life unharmed.

As water technology becomes more and more advanced and refined, 'everything becomes possible'. Our technological options provide us with so many options. But at the same time, to have gained access to that power requires us to be transparent in our aims; to ensure benevolent use of that power. Water is a primary substance of life; hence, we influence life directly in our interaction. Our own internal story about water invisibly governs our decisions and actions with water. We need to spend time and attention in reflecting upon the future of water. As water has become a vital topic of the perilous global sustainability crises, this need becomes ever more acute. As technology gets better, we need to upgrade our moral, emotional, spiritual professional selves as well. My findings lead me to believe that this marks a transition from water professionals to water stewards.

However, before discussing what is not working well in our current state, we must recognise that the industrial revolution was not carried out of spite and evil and the results are not solely disastrous. On the contrary, most people have benefitted from fantastic progress. The heavy hand of a plethora of diseases has been lifted. The world has experienced an incredible increase in wealth. The internet has provided almost free availability of an ever-increasing domain of shared knowledge. The decrease in human violence has continued. But underneath this gushing forward of comforts and benefits, a discomforting sound has appeared. And this discomforting sound has grown loud enough for a new disorientation. There is a sense that something new and different is underway. 'Smart' or 'intelligence' has to be supplemented by something of a different dimension. Something in the direction of 'wisdom', 'the best of humanity', 'a caring respect for nature' and 'poetic beauty'.

It appears increasingly evident that a change is in its waking, and that this transformation happens everywhere. I wanted to understand the change, the drive for change and how it could be applicable to our relationship with water. This book is a manifestation of that understanding.

# Chapter 1

# Aspiring to a new story

In this chapter, I will investigate and attempt to crystallise the drive towards a new way of relating to water. During the process of writing this book, I have developed my capability to speak from this drive and about this drive. And I find that a surprising number of people themselves sense a similar drive. For some, it is strong and has weight and volume in their everyday inner life. For others, it is only a faint sound that springs up under circumstances of long slow conversations. It is as if there is a secret almost silent trickle inside. For most, however, the conversation of this theme is difficult to carry out. It is awkward, confusing, messy and unclarified, words are missing. It feels like trying to access an unresponding part of the brain. Charles Eisenstein speaks of the state we are in as 'a place between stories' (Eisenstein, 2015). Our old story organising our intuitive understanding of everything is breaking down while at the same time, a new story is not yet available to us in full. We are like the polar bears we see in photos standing on slates of melting ice with no new ice slate to move to.

The interpretation of the drive is carried out by finding metaphors, concepts and imaginative ideas that point in that direction (Figure 2). It is a work of trying to capture something elusive. The drive is expressed differently in each of us, but there are also strong similarities. During this work I repeatedly come across similar patterns from all over the world. Hence the aim here is a recognition and exploration of this inner landscape more than it is a scientific excavation of a new principle. It is asking: how does this drive feel in me? What does it mean to me and what feels true about it?

© IWA Publishing 2020. Water Stewardship
Author: Pernille Ingildsen
doi: 10.2166/9781789060331_0001

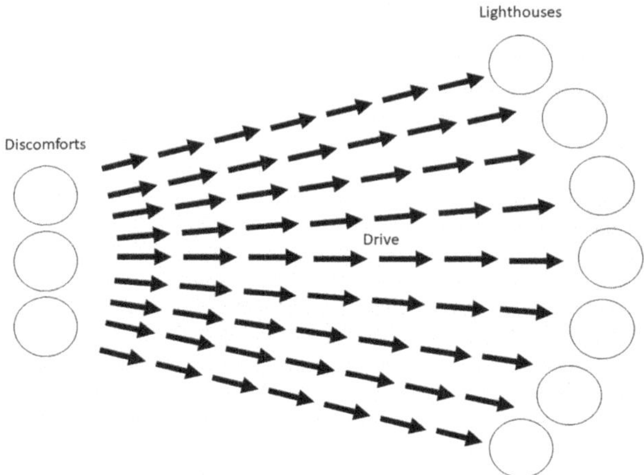



**Figure 2** Illustration of the concept of discomforts and lighthouse stories determining our inner drive.

It has been challenging to find a method to describe this systematically. An important step has been to isolate and investigate the primary discomforts in how we operate now. Step two in the process has been looking for ideas and accounts that have a soothing effect on the pain of the discomfort; an understanding that can truly respond to the discomforts; an understanding that contains an inspiring spark of something new. Understanding the drive from the discomforts to the new is a process of searching backwards and inwards to understand the inner aspirational drive. The negative discomforts and the positive possibilities show the direction. And they provide the fuel forward towards a new story. It is almost as if the future has lighthouses beaconing in the horizon.

After understanding the critical driving forces begins the work of imagining and visioning. How does a radically different future with a healthier relationship with water in a more beautiful world look and feel? What would be some of the governing principles? We must approach this thought experiment with an open mind, an open heart and an open will. To experiment means we need to lose our grip on the current story and attempt a different footing.

*'To dare is to lose one's footing momentarily. Not to dare is to lose oneself.'*
Soren Kierkegaard

**Your reflections:** If the idea of discomforts and lighthouse stories make sense to you, how would you describe your discomforts and which lighthouse stories provide you with a soothing sense of hope?

I have found three essential discomforts that encompass a kind of skeleton description of the current trouble. These are feelings of desecration, apathy and banality. In the following, I will delve into a deeper understanding of what these feelings can mean.

## THE DISCOMFORTABLE FEELING OF DESECRATION

The feeling of desecration links to what Gustaf Olsson tried to convey when speaking about water being holy or sacred. By desecration I mean the continued acts of not keeping 'the sacred' sacred. We can observe this lack of sacred outlook and actions when we pollute water or when we view water first as a resource for our consumption, a resource to be treated and traded as we see fit without thought for the role and meaning water has in the whole picture.

Inside me, it looks like a gap from where we are to a 'sacred line'. On one side of the sacred line we see ourselves as 'being an integral part of nature' on the other we have 'a human-centred exploitative look on nature'. A special quality of dedicated care and attention is required to stay on the honourable side of that line. The concern of this discomfort is that we have moved so far away from the line that it is difficult even to see the line from where we stand. And, we left the line so long ago that we have almost forgotten it. This 'dreamland border' is quickly forgotten when living in the 'ordinary world' of everyday life with seemingly abundant water conveniently available.

However, regardless of our forgetfulness, wherever we go there is always this country behind the curtain. We may be able almost to forget it, to almost ignore it, to almost repress it – but always only almost.

> *'Behind the cotton wool is hidden a pattern; that we – I mean all human beings – are connected with this; that the whole world is a work of art; that we are parts of the work of art. Hamlet or a Beethoven quartet is the truth about this vast mass that we call the world.'*
>
> Virginia Woolf (1976)

Acting as if the land on the other side of this line does or cannot exist exhausts me, and it is becoming increasingly clear to me that it is a disconnection. There is in this disconnection a kind of dishonouring. And one thing is that it is hurting our environment but it also hurts ourselves in the stories we tell about ourselves and the stories we cut ourselves off from and thereby limit our potential. And by this realisation, we see an essential misunderstanding about our distinction between ourselves and the so-called environment. It is not that I feel that there is no difference between me and the environment and that 'it is all one', but instead it becomes more and more clear that what we do outwardly we also do inwardly to ourselves.

> *'People who dream when they sleep at night know of a special kind of happiness which the world of the day holds not, a placid ecstasy, and ease of heart, that are like honey*

*on the tongue. They also know that the real glory of dreams lies in their atmosphere of unlimited freedom.'*

Karen Blixen (1937)

If there were to be a true heart-felt honouring of water, we would be crying about what is happening. We would be crying for the oil spills in the Mexican Gulf or the devastating oil spills during decades in one of the most important wetlands in the world, the Niger Delta in Nigeria. We would be crying over toxic chemicals of pesticides, medicine and beauty products found in water everywhere. We would cry about the Samarco mine tragedy in 2015 (see, for example, Gormezano *et al.*, 2016) where toxic mud poured out after a burst mine dam; and then travelled 500 km by the river of Doce to the Atlantic Ocean, colouring everything red-orange on its way – leaving a trail of toxic heavy metals and other substances that only centuries will erase. We would cry over the taste of chlorine in our drinking water. We would cry about the animals whose habitat is defiled by plastic debris everywhere. We would cry over the impoverished state of our streams flooded by wastewater and a constant seeping in of fertilisers and pesticides. We would cry with the many silly, invaluable products we produce with the consumption of water for the sake of a short, shallow gratification. If we genuinely believed that water was sacred, this would be a world of grief.

This suppressed grief and despair is described in all its horror by Eve Ensler (2014). With this text, suddenly, what we have unknowingly ignored, repressed and placed in our blind spot is made visible, and the pain is almost tangible.

*'I began to see my body like a thing, a thing that could move fast, like a thing that could accomplish other things, many things, all at once. I began to see my body like an iPad or a car. I would drive it and demand things from it. It had no limits. It was invincible. It was to be conquered and mastered like the Earth herself. I didn't heed it; no, I organized it and I directed it. I didn't have patience for my body; I snapped it into shape. I was greedy. I took more than my body had to offer. If I was tired, I drank more espressos. If I was afraid, I went to more dangerous places. [...]*

*Then I got cancer – or I found out I had cancer. It arrived like a speeding bird smashing into a windowpane. Suddenly, I had a body, a body that was pricked and poked and punctured, a body that was cut wide open, a body that had organs removed and transported and rearranged and reconstructed, a body that was scanned and had tubes shoved down it, a body that was burning from chemicals. Cancer exploded the wall of my disconnection. I suddenly understood that the crisis in my body was the crisis in the world, and it wasn't happening later, it was happening now.*

*Suddenly, my cancer was a cancer that was everywhere, the cancer of cruelty, the cancer of greed, the cancer that gets inside people who live down the streets from chemical plants – and they're usually poor – the cancer inside the coal miner's lungs, the cancer of stress for not achieving enough, the cancer of buried trauma, the cancer in caged chickens and polluted fish, the cancer in women's uteruses from being raped, the cancer that is everywhere from our carelessness. [...]*

*I know that everything is connected, and the scar that runs the length of my torso is the markings of the earthquake. And I am there with the three million in the streets of Port-au-Prince. And the fire that burned in me on day three through six of chemo is the fire that is burning in the forests of the world. I know that the abscess that grew around my wound after the operation, the 16 ounces of puss, is the contaminated Gulf of Mexico, and there were oil-drenched pelicans inside me and dead floating fish. And the catheters they shoved into me without proper medication made me scream out the way the Earth cries out from the drilling.'*

Eve Ensler

I put this pain here in the centre of us to help us remember where we start and why we need this journey. Some may find the description overly dramatic, but even so I believe that for all, there is a recognition. I could fill pages documenting the state of water everywhere. Although the disaster stories that arrive through broadcasted media are exceptionally horrendous, they describe a widespread trend. This is the direction we are heading in everywhere. Everything is a matter of degree from the best state of water to the worst. But the best state, which was not so long ago the natural state everywhere, is disappearing. 'Oh, here it is as it should be – still'. There are problems everywhere. There is no secret remote refuge where the water is pristine.

That is a sad thought, and the feeling of this thought is an essential part of the discomfort with where we are. We are not where we collectively want to be with water. It is so far from our hopes and aspirations and what feels right. It becomes increasingly strenuous even to visit the original situation in our imagination. We struggle already in these years to remember the old natural state of nature. The struggle to remember will only get harder with each passing year. Already now we need to investigate historical records of different kinds to try to piece together an image of the original nature.

---

**Your reflections:** In our modern world water is not seen as sacred. Imagine you could make yourself see water as sacred. How would that change what you do professionally and privately? Would it feel different when approaching issues that have to do with water? Would it change the feeling of the water in you?

---

## LIVING WITH A SACRED LOOK AT NATURE

The documentary 'Aluna' (Eirera, 2012) features BBC journalist Alan Eirera who in 1990 made a movie about the Kogis. The Colombian indigenous Kogi people live in the area of Sierra Nevada de Santa Marta. Here they follow their indigenous way of life and traditions. The Kogis asked Alan Ereira in 2012 to make a new movie. The Kogis stated that their purpose was to send a message of warning to 'little brother' (their word for modern man). To make him aware that earth and nature are suffering and dying.

I watched the documentary twice. The first time my primary feeling was confusion and frustration. And the frustration was shared with everybody in the documentary. The communication gap between Alan Eirera and his film crew on one side and the Kogis on the other side is so wide and deep that it appeared almost impossible to communicate across it. Not so much due to language but due to widely differing mind worlds. Hence the documentary is not only about the environmental message the Kogis were trying to deliver; even more striking is the frustrating gap hindering their common understanding. Something evident and clear from an indigenous perspective is entirely invisible and incomprehensible from a western perspective.

The movie starts with a summary of the creation story of the Kogis:

> *'In the beginning, there was nothing. All was darkness. There was only the mother. She was Aluna. She was pure thought without form. She began to think. The mother conceived the world in the darkness. She conceived us as ideas. As we think out a house before we begin to make it. She spun the thread. Spinning us all in the story. Creating us in thought. And then came the light. And the world was real.'*
>
> Quote from the movie Aluna

The Kogis find it evident that critical networks connect special places. And they find it obvious that modern man's activity is destroying these 'special places'; and that the activities are deeply detrimental and catastrophic. According to the Kogis, a black thread of thought connects the unique places. To make the connections visible to 'little brother' the Kogis embark on a pedagogical mission with Alan Eirera. Their best way to explain this is to take a thread of gold (actual real gold) and physically connect the special places to make the connection visible. Hence, they bring a 400 km long gold thread to the start and drag it from there through a number of specific sites at the mouths of rivers.

> *'We, the mamas [enlightened priests of the Kogis], know that there are special sites and they are threaded together. We're explaining this to our brothers across the sea and showing the connections between places. … The lakes that are found high in the mountains are linked to the sea. Bubbles appear in the water, and the bubbles make the connection, that is how we communicate with the lakes and the sea. The water speaks.'*
>
> Kogi Mama (Aluna)

Again and again the Kogis fail to explain the problem in a way that Alan Eirera comprehends. But after many attempts, Alan finally finds a professor who moves the understanding forward. Professor Jonathan Baille says:

> *'In conservation we have a problem because we tend to conserve an area without much thought to the special places in a geological time-scale. So, for species, there are areas that are like a refuge, which, through geologic time, have been extremely important for them to survive. And in the short term, we may remove these areas, and there may be no*

*big effect, but in the long term, through history, the species can't persist when those special zones have disappeared.'*

Jonathan Baille (Aluna)

The Kogis nod eagerly, happy and relieved to find some common ground finally.

*'We have to better understand connectivity, and right now, we have a very basic understanding of how things interact and affect each other. And I believe that this is essential for our future's security, to understand these special sites, to ensure that they are conserved.'*

Jonathan Baille (Aluna)

The movie is a pain to watch. On several frustrated occasions, Alan tells the Kogis in a very direct form that he finds it very difficult to see their point in the whole film project. It is a pain for the Kogis, who are trying to explain what is self-evident for them. Like a fish asked to explain how it is to swim in water. And in the end, when we finally get enlightened, it is a painful understanding.

The understanding is that places are connected, and specific points bear great importance. At some point we may be able to understand this importance from a scientific point of view; by investigating the places in the perspectives of ecology, water cycles, diversity, evolution, geology etc. But more likely we may experience the consequence by accidentally destroying these places.

From a Kogi point of view, this is however only half of the story. There is also an important spiritual dimension. For the Kogis, it is not only a question of securing physical livelihood in the form of functional ecosystems. There is more to it than the basic scientific mapping of relevant nutritional cycles, water cycles or habitats. In their eyes 'little brother', western man, is acting like a drunk man on a large scale, not aware of the havoc he is causing as he stumbles around.

The Australian Aboriginal idea of song lines has a similar feel to it. It has a similar sense of a world hidden beneath what we can see with our eyes. When I visited the University of Queensland in Australia, I was so looking forward and even excited to read the stories from Aboriginal mythology. But to my disappointment, the stories left me confused and unable to grasp the point of them. The stories felt almost incomprehensible. Recently I picked up the mantle again of trying to understand. As I read about song lines, I suddenly gained some ground of understanding. The stories seemed 'pointless' because 'the point' was something different from what I expected – it was not about heroes and villains.

Songlines, also called dreaming tracks, criss-cross all of Australia. A songline is a track over land connecting distinct places such as waterholes, hills, rock paintings and other landscape markings. A songline may cross and connect different groups of people with different languages. The songlines can be sung and describe the story of trajectories of ancient ancestors who marked the country in the Dreamtime (the time of creation according to Aboriginal mythology). The Aborigines see themselves as custodians of the country. Each tribe has the responsibility for keeping the part of the songlines that crosses their territory alive. Bruce Chatwin, who in 1987 brought knowledge of songlines to the public, wrote about the Russian Arkady who taught him about songlines. Arkady learned about the 'song lines', the 'Dreaming-tracks', the 'Way of the Law' or the 'Footprints of the Ancestors' when he was a school teacher:

*'Aboriginal Creation myths tell of the legendary totemic beings who had wandered over the continent in the Dreamtime, singing out the name of everything that crossed their path – birds, animals, plants, rocks, waterholes – and so singing the world into existence.'*

Bruce Chatwin (1987)

Arkady was asked to help with drawing the line for the railway going straight south–north from Alice to Darwin. Arkady was asked to help avoid going through too many sacred places: 'Well, if you look at it their way', he grinned, 'the whole of bloody Australia's a sacred site.'

Laine Cunningham, who travelled in Australia to collect some of the song line stories in her book Seven Sisters (Cunningham, 2013), describes it like this:

*'According to Australia's ancient cultures, all creatures and things emerged from the Dreamtime. The Dreaming is not just a collection of lore or a long-ago time; it is a living energy that flows constantly through the universe. It is then and now, divine and human, spirit and law. It teaches us how to survive in a harsh world and how to thrive in our souls.*

*Most clans conceived of a creation in which Earth already existed. Ancestors rose out of the ground and descended from the sky. Wherever their feet pushed up mounds, mountains arose; wherever the ancestors fought, the ground was trampled flat.*

*Tribal members can still 'read' the land by walking a story's path, its songline. In this way, the people were connected to the land. The largest song lines, epic stories of ancestors who ranged far across the continent, connected different tribes. When an ancestor crossed into new territory, the next part of the story belonged to the neighbouring group. The entire song line could only be recited when all the tribes had gathered. Relationships between neighbours were therefore automatically – and spiritually – strengthened.'*

Laine Cunningham, author

'But!' one may object, 'I am no longer indigenous. I belong to a different scientific culture. We left the indigenous living and mindset many generations ago. You surely don't want me to pretend I have this type of bond to nature?' No, perhaps not. But there is something in this that rings true and rings like something lost. Indigenous people still keep contact to their mythology; they keep it alive. And it is as if they are guarding something important for all of us – a living thread back to a more sacred look on life. In Australia, the Aboriginals have maintained their society and livelihood sustainably for millennia in the difficult Australian terrain. This way of thinking has served an essential purpose of ensuring the continued way the aboriginal culture has thrived and survived. The idea of songlines is both elegant and enjoyable to think around. There is something in it that passes the platonic test of being good, true and beautiful.

If I look at our current modern world through the eyes of indigenous eyes, I am overwhelmed by the particular kind of poverty we have created in our society. When I walk outside my house, my landmarks are gas stations and supermarkets and natural places like forests and lakes have watered down, stumped and numbed.

*'There are no unsacred places, there are only sacred places and desecrated places.'*
Wendell Berry (2005)

Wendell Berry points at an important point, that all places are sacred until treated with disrespect. Implicitly they can become sacred again when the violation is made right, when we start acting 'sacredly' towards these beautiful natural living phenomena. We have to find a way to bring places back to life in us, to re-understand and reconnect with where we live. In the same way, there is a longing to find a way back for water; to gain a heart for water; to reconnect to water mythology and to treat water with greater respect and eventually sacredly.

**Your reflections:** Indigenous people (aboriginal or native people) carry with them a culture from before industrialisation. Many have existed sustainably for centuries. They have a closer bond to nature. What kind of things do you think we could learn from them in terms of water? Might they contribute with principles or ways of bonding with nature that could be helpful where we are today?

# THE DISCOMFORTABLE FEELING OF APATHY

I know the feeling of apathy inside me. There are degrees of apathy, and it feels like a wind I stride against in myself. Like a tiring resistance of will.

But when I learned of the origins of the word apathy in an interview with ecologist Joanna Macy, something fell into place; I understood the source of the apathy.

*'I realised the etymology of the word was a reflection of what was so. [early 17th century: from French apathie, via Latin from Greek apatheia, from apathēs "without feeling", from a- "without" plus pathos "suffering."] It was not that people didn't care or didn't know, but that people were afraid to suffer. It was the refusal or the incapacity to suffer.'*
Joanna Macy (Jamail, 2017)

This description made me understand the reason for apathy and the reasons for its occasional disappearance. I know both the avoidance of suffering in me and the despair when the bulwarks of avoidance are flooded. I have always felt this as a lack of strength in me, that I should be so susceptible to pain and despair and unable to stay invariably positive and without sorrow.

But perhaps it is true that our bulwarks of avoidance against becoming touched by suffering are the locus of our apathy. In a water perspective, our avoidance of feeling suffering perpetuates suffering. It causes our actions to lead to water pollution, droughts and flooding – or even if we try to 'do good', the lack of feeling of suffering blunts our actions.

Letting in suffering is hard. It hurts. It requires courage. Hence its repression is often our automatic subconscious standard response to oncoming suffering. The avoidance keeps us comfortable – at least in the short time-scale. But for happiness and bliss to return, we must strive to live wholeheartedly. That means being aware of what is happening. And regardless of what positive or negative realisation presents itself, to take that in wholeheartedly.

Hence if we have chosen to work with water in our lives then let us do so wholeheartedly. That is the challenge we must meet. In the same way that we need those who work with judicial matters to strive in their actions to move us towards justice wholeheartedly. And those that work with healthcare to move us towards health wholeheartedly. And those that work with education to move us towards competence wholeheartedly and so forth. When we look around, it is clear that if we are not going to speak and act for the liveliness of water who is? If not us in precisely this place and this time? Who else is going to ensure good and healthy water for people and nature here?

Being a water professional is a big responsibility; just as ensuring justice, medicine and teaching is. To deliver on this responsibility is not a 'walk in the park'. And though I occasionally see heart-based approaches in utilities, I more often see a lack of wholeheartedness; a lack of paying attention and a lack of courage; of playing it safe, of taking short-cuts, and of not going the extra mile, to deeply understand the problems at hand. I see, again and again, water managers, politicians and professionals holding up legislation or economic restraints as if they were solid barriers; barriers against doing the right thing or against thinking for ourselves to understand this situation deeper. And for sure, there are plenty of barriers. But the barriers do not legitimate giving up on doing what is needed for protection or restoration. We need to search for ways to improve our mode of water operation to become sustainable and aligned with our higher ideals.

The ignorant progress-optimistic innocence that was the dominant mindset in the '60s, '70s and perhaps still in the '80s, is gone forever. We know better, and we can see or imagine the suffering of our actions with no great difficulty. If the consequence of this tendency to avoid sensing the suffering is apathy, then it is not only an irresponsible stanza but also a very unhappy solution for us individually as well as collectively. As the days count on, the physical world is made relentlessly worse by processes of unsustainable behaviour. And individually, we unconsciously sense stuckness in this unhappy state of apathy.

The uncomfortable question then is: what to do with this suffering feeling, when neither repression nor being overwhelmed by it are good solutions? Finding an answer to this question is key to the legacy we will leave behind us.

I recognise a truthfulness in this quote from Rainer Marie Rilke. He manages to describe the feeling of suffering, the solution and the result in one short quote:

*'It seems to me that almost all our sadnesses are moments of tension, which we feel as paralysis because we no longer hear our astonished emotions living. Because we are alone with the unfamiliar presence that has entered us; because everything we trust and are used to is for a moment taken away from us; because we stand in the midst of a transition where we cannot remain standing. That is why the sadness passes: the new presence inside us, the presence that has been added, has entered our heart, has gone into its innermost chamber and is no longer even there, - is already in our bloodstream. And we don't know what it was. We could easily be made to believe that nothing happened, and yet we have changed, as a house that a guest has entered changes. We can't say who has come, perhaps we will never know, but many signs indicate that the future enters us in this way in order to be transformed in us, long before it happens. And that is why it is so important to be solitary and attentive when one is sad: because the seemingly uneventful and motionless moment when our future steps into us is so much closer to life than that other loud and accidental point of time when it happens to us as if from outside. The quieter we are, the more patient and open we are in our sadnesses, the more deeply and serenely the new presence can enter us, and the more we can make it our own, the more it becomes our fate.'*

Rainer Marie Rilke (1929)

Equipped with this understanding, I find it possible to have a more relaxed and mindful reaction to these hours of dark suffering. To go through these suffering processes, these doors, not as inconveniences, but as moments of transformation. Repeated, slow changes in our mindset and 'heartset'. When carried through mindfully, these painful periods present practice of a softening of the heart and a way out of apathy. Be still.

> **Your reflections:** What is your relationship to suffering? How do you empathise with other people? Can you sense or imagine suffering in animals? In plants? In all living organisms? How do you feel about the environment as it is outside your home? How well do you know it?

## THE DISCOMFORTABLE FEELING OF BANALITY

Elisabeth Minnich highlights the extremity of living a life of banality. She comes from a different starting point, but her findings carry relevance here. Minnich is a student of the famous philosopher and researcher Hanna Arendt, best known for the idea of 'the banality of evil'. The concept of 'the banality of evil' was developed while following the lawsuit against Erich Eichman. Erich Eichman was one of Nazi-Germany's leading architects. Hanna Arendt's finding was that Erich Eichman was not a uniquely evil human being, but rather quite ordinary and 'banal'. She found that when the process of Nazism was first in motion, banal people were very useful in implementing and perpetuating it. Her point was that what had happened in Germany could have happened in many other places – given certain conditions. This, of course, in itself is a valuable insight. However, her student Elisabeth Minnich turns the concept of 'the banality of evil' a bit on its head:

> *'When Arendt took me with her to public discussions of her book, which were often acrimonious and painful, what struck me was, first, that "the banality of evil" was deeply troubling to people, as it seemed to trivialise enormous wrongdoing and suffering. It occurred to me that, had Arendt spoken of "the evil of banality", what she meant might have made more sense. The thought has stayed with me that we need to comprehend how banality – superficiality in its differing modes –works to enable its apparent opposite, the great drama of horrific harm-doing.'*
>
> Elizabeth Minnich (Ballowe, 2018)

In her work, Elizabeth Minnich differentiates between two types of evil: intensive and extensive evil. The 'banality'-hypothesis of Eichmann is not useful when understanding Hitler; he is not banal. Intensive evil is about the evil psychopaths' living among us, who set out to harm us on purpose for their own benefit. Extensive evil, on the other hand, is more commonplace, and we are all perpetrators of extensive evil due to our thoughtlessness and our unreflected ways of operating which is part of our culture. Extensive evil is the repeated pattern of what we accept in our culture that upon closer scrutiny turns out to be wrong by our standards and values. It is a normalisation of things that we, in our hearts, know to be evil. If we were individually asked to decide about such actions, we would not find them to be ok. But we are not asked to decide or carry many of the actions out personally; they are already embedded 'in the system' or in our invisible culture.

Elizabeth Minnich explains:

> *'It is* intensive *evil for a pyromaniac to launch a forest fire, or for a criminal cast out of a village to sneak back and poison its wells. It is* extensive *evil when the pillars of society and a whole economic order accepts as mainstream that pollution of whatever sort and consequence is understood to be a mere bother that must not derail profits, economic growth, freedom from governmental interference. When careerism, greed, status-seeking are rewarded for work that is horrifically harmful to large numbers of people, even to a viable future for all, we have a classic case of extensive evil. It is not perpetrated by monsters; it is perpetrated by those who think no further than how to play and try to win in the present game, by the dominant rules.'*
>
> Elizabeth Minnich (Ballowe, 2018)

When we look at the suffering and destruction that take place in our shared sustainability crisis, extensive evil plays the main role. Intensive evil can hardly be played out on such a global scale, without the help of extensive evil. Our habits, culture and mindset blind us so that we cannot see what part of our everyday life is causing extensive evil. We are both 'victims' and perpetrators of our own 'evil of banality' – and often we must take the role of rescuer; perpetuating the eternal drama triangle. Our world interpretation and our understanding of our role in our world are too oversimplified, too banal. There is a heavy shadow side in our lives that we continue to deny. It has shown to be a very difficult pattern to break: 'I will not change until the mainstream changes'.

The banality prevents us from understanding the deeper impacts when we extract too much water or lead out poorly treated wastewater in recipients – not looking into the consequences and not understanding – eyes wide shut.

> **Your reflections:** Do you know the feeling of shutting yourself off when you have to do something that does not feel quite right? Do you recognise having experienced aha-moments that have changed your perception of something, so that something that you did before that you will not do again? If you look out at life in general, are there things that you need we will not do ten, twenty or fifty years from now on moral grounds?

The three feelings of desecration, apathy and banality capture the essence of a gap in the integrity of our emotional landscape towards water. Acknowledging these feelings has been difficult on more levels. It was difficult to understand that they were the ailment. It was difficult to accept my complicity. It took time to dig into the research on these topics. The desecration causes suffering that I am not letting in, causing apathy. The banality of my understanding of myself and the world and the local consequences of our water use causes my apathy to continue and my actions to be superficial. So, a significant shift is needed.

It is as if I need to squeeze myself through something and from there feel my way through what soothes and balms the pain of these feelings. Upon hours and weeks of contemplation, I found that it had to do with the heart. It had to do with not understanding or even listening to my own heart. At the onset I hardly knew what I meant by that. Therefore it has been step-by-step learning to lead a more wholehearted life. I have not reached my golden goal yet, but I have learned a lot. I have learned something that shifted my being. It had to do with deeper understanding, wholeheartedness and attending to the pain of these discomforts.

Various lighthouse stories have captured a sense of different possible futures. Ideas, concept and stories that enrich and direct an evolving mindset in the water journey ahead. In the following I will provide an understanding of the role of the heart on three interlinked levels: the personal, the local and the global. Together

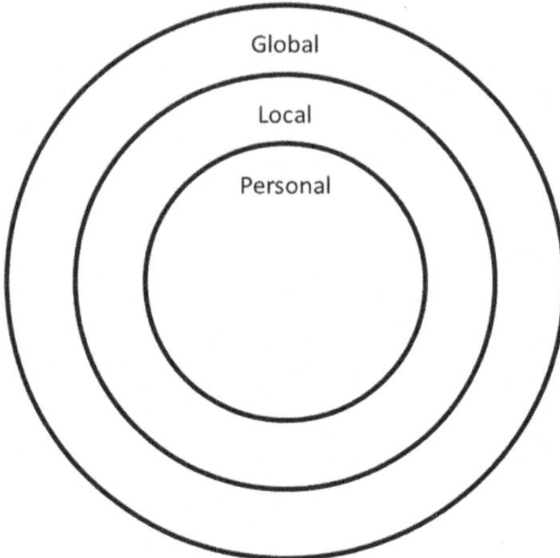


**Figure 3** A way to illustrate a more holistic view is to consider the three layers of stories we need. The personal, the local and the global. All three levels must be comprehended and addressed in our work moving forward.

they form a kind of integrated understanding of how to apply more heart in our work as water stewards (See Figure 3).

## THE ROLE OF THE HEART IN THE PERSONAL

In H. C. Andersen's story 'The Nightingale' (Andersen, 1843), we are introduced to an emperor with a beautiful garden and a beautiful forest.

> *'Those who travelled beyond its limits knew that there was a noble forest, with lofty trees, sloping down to the deep blue sea, and the great ships sailed under the shadow of its branches. In one of these trees lived a nightingale, who sang so beautifully that even the poor fishermen, who had so many other things to do, would stop and listen.'*
>
> H. C. Andersen (1843)

When the emperor read about the bird in a book, he wanted to hear it sing. So he summoned his court to bring it to him. After some difficulties, the court caught the bird and he finally heard it singing. The sound of the song stirred the heart of the emperor so much as to bring tears to his eyes. The bird brought him so much happiness that he installed the nightingale in a cage in the palace.

One day, however, the emperor received a gift from Japan.

> *'It was a work of art contained in a casket, an artificial nightingale made to look like a living one, and covered all over with diamonds, rubies, and sapphires. As soon as the artificial bird was wound up, it could sing like the real one, and could move its tail up and down, which sparkled with silver and gold.'*
>
> H. C. Andersen (1843)

The artificial bird created lots of amusement and attention, and during all this commotion, the real nightingale took her flight home to the noble forest.

> ' "But we have the best bird after all", said one, and then they would have the bird sing again, although it was the thirty-fourth time they had listened to the same piece.'
>
> H. C. Andersen (1843)

For a while, the artificial bird created much happiness. However, as with all mechanical things there came the point in time where it was almost worn out and could only be wound up once a year.

Five years later, the beloved emperor fell seriously ill.

> 'A window stood open, and the moon shone in upon the emperor and the artificial bird. The poor emperor, finding he could scarcely breathe with a strange weight on his chest, opened his eyes, and saw Death sitting there. He had put on the emperor's golden crown, and held in one hand his sword of state, and in the other his beautiful banner.'
>
> H. C. Andersen (1843)

The people who usually were around him had left to vie for the future emperor. In his misery, however, the real nightingale paid him a musical visit. With its song, the bird charmed death away and restored the emperor to health. As he wanted to thank her, she replied,

> ' "You have already rewarded me," said the nightingale. "I shall never forget that I drew tears from your eyes the first time I sang to you. These are the jewels that rejoice a singer's heart. But now sleep, and grow strong and well again. I will sing to you again."
>
> "You must always remain with me," said the emperor. "You shall sing only when it pleases you; and I will break the artificial bird into a thousand pieces."
>
> "No; do not do that," replied the nightingale; "the bird did very well as long as it could. Keep it here still. I cannot live in the palace and build my nest; but let me come when I like. I will sit on a bough outside your window, in the evening, and sing to you, so that you may be happy, and have thoughts full of joy." '
>
> H. C. Andersen (1843)

The heart is a metaphor for intention, care and emotion. In utilities, the heart shows up many places. Most significantly, it turns up in the choice of career. Many water professionals in the water sector see their work as a dedicated service. Unmistakably there is an element of 'calling' in the grit that is exhibited by many water professionals. Yet at the same time, the heart in the workplace is unacknowledged – a topic in the realm of the unprofessional, the idealistic or naive. The heart is not discussed in the collective space or included in an individual practice – it is hidden and unappreciated behind the scenes.

Through the last years, I have studied and experimented with 'the heart in relationship to water'. I have found a multitude of reasons why it is relevant to dedicate time and attention to explore this issue. 'The heart' is a strange self-entangled topic to study, where the long honoured concept of objectivity is 'out of order'.

Moving in the direction of the heart has occasionally felt like moving against my own normal work culture; a technical culture where science and objectivity is the golden standard. But it has been worth the effort and more often than not the two perspectives have supplemented each other.

From a sustainability perspective, we need somehow to upgrade our standards and practices. Something fundamental and heart-related seems missing in our internal 'manual of operation'. I hypothesise that this has to do with depth. The commodification of water has caused something to be lost – something of a 'sacred' quality.

When looking at our work as water professionals through the viewpoint of the heart, important questions arise. Questions like: What is the meaning of my work – or rather: What is the purpose of my Work. The distinction being that *work* is the mundane day-to-day work, projects and tasks, while the *capital-W Work* is the existential part of what we do – related to one's life purpose. How do I understand our history and culture of the Work with water as per where we are today? What does it mean for me to fully face the current state of water sustainability crises? How can I embrace that responsibility moving into the future? How can I transform the untold aspirations energising the Work into actions that are satisfying on a deeper level? How can I at the same time produce positive change for water, people and ecosystems?

You may ask: Why work from a heart hypothesis?

The answer is: because I feel this constant calling of something better, more authentic and more beautiful. Something possible, but still unaccomplished. There is undoubtedly potential for a higher level of grace. For a while, I thought this was about 'austere' sustainability: reducing water footprint, food consumption, travelling, moving, heating, cooling, etc. to be content with less. Yet, though the unsustainability crises are large and growing, I find it hard to generate a steady impetus for a goal of ever reducing my footprint. It is as if the end of that road is less lively; feels less alive. And this cannot be true.

Our global water sector makes progress on more parameters, but the development seems too slow to bend our path enough to avoid crises. On some parameters, the sustainability crises even seem to be becoming more entrenched. Instead, I hope and think that if we can advance to the next rung of the developmental ladder then the trouble of fitting all our feet on the space of earth will be resolved. I believe it is possible to achieve that without a feeling of loss but rather with a feeling of gain. I hope that when we find a way to live within a sustainable culture, we will stand on healthier feet and that our current neediness will not continue to spoil the water, nature and the earth around us. It simply seems as within reach to find a more mature way of sustainability. A way that does not have the feeling of 'disciplined austerity', but instead feels vibrant, alive and attuned.

Thus, this 'search project' is for a new story about water and life. It is born out of an inner urge to find a way that makes me able to lead my life in consistence with both my heart, my intelligence and my higher aspirations. To have a new pathway emerge towards a more graceful relationship with water by strengthening the connection to the heart.

The Danish researchers Steen Hildebrandt and Michael Stubberup express a similar belief:

*'One cannot practice sustainable leadership unless it is grounded in a profound anchorage in the self. Sustainable leadership is generated by our ability to read feelings as sounding boards for dreams, intentions, decisions and actions.'*

Steen Hildebrandt and Michael Stubberup (2012)

Therefore I am looking to experiment with routes to life with the heart in the centre. First, I wrote the sentence as '... back to life with my heart in the centre', then upon reflection, I was wondering is it instead '... a way forward to a life with my heart in the centre'? I will let this question stand open but share the doubt and wondering. But I tend to believe that it is a road forward, that we can integrate heart and head in new rewarding ways. Regardless, my experience is that finding such a way is easier said than done. It is as if the muscle for doing this is not well practised; something different must be introduced for the heart to take centre stage or for establishing a strong connection between head and heart.

At the beginning of my experiments, I was very concerned that a heart approach would mean a departure from the road of natural science. A road that I, as an engineer, had followed for decades and from which I had seen quite effective results. However, I have found that science and the heart are not per se in conflict. By applying 'looking through a heart lens', the field of water looks different and different questions come up. Projects become defined differently, different relationships are formed, and different results ensue. Still, the scientific way of thinking and learning is essential and extremely helpful. Hence, I can strongly disagree with my first fearful misconception: a heart approach is not a departure from rationality and science. Instead it is a step further towards the original source of motivation and meaning.

In his book 'Finite and Infinite Games', James Carse explains why a new mindset is not meant to be an abandonment of the older mindset. He posits that culture develops as people change the rules by making deviations that build upon what came before:

*'There are variations in quality of deviation; not all divergence from the past is culturally significant. Any attempt to vary from the past in such a way as to cut the past off, causing it to be forgotten, has little cultural importance. Greater significance attaches to those variations that bring the tradition into view a new way.'*

James Carse (1986)

Hence, the point is not to 'cut the past off'. We need to bring our past with us – to integrate the past in the future. Integration needs to be done both in the form of building upon successful results and in the form of acknowledging the damage done.

This integration cannot be done superficially. It is paramount that this integrating step forward is based on a wholehearted approach. To accomplish this, we have to delve into and appreciate the complexity of this integration. We often work towards reducing complexity, attempting to reach decisions based on a simple foundation of logical and rational – but often reductionistic – argumentation. But in this new world, we have to proceed with much more care and attention or we will lose something important through our simplification. We must train ourselves to become able to comprehend a more full and holistic understanding of the systems at hand. If anything, this new world is about integration and inclusion – not reduction and exclusion.

Along the same lines, Einstein is quoted to have said:

*'Creating a new theory is not like destroying an old barn and erecting a skyscraper in its place. It is rather like climbing a mountain, gaining new and wider views, discovering unexpected connections between our starting points and its rich environment. But the point from which we started out still exists and can be seen, although it appears smaller and forms a tiny part of our broad view gained by the mastery of the obstacles on our adventurous way up.'*

Albert Einstein

Hence, our 'smaller views' are not necessarily deeply wrong, they are just small. We need to broaden and expand; to reach some higher form of knowledge – access and insight into universal wisdom perhaps. So, the role of the heart is not to be 'whimsical', but rather to be a solid base of impetus to learn deeper, to prepare better, to achieve a more beautiful attunement of our actions with water as the lifeblood of the biosphere. The role is to 'sense the universal wisdom' and help translate it to knowledge to inform our actions.

This manoeuvre requires a 'change of heart'. Or more precisely, it requires a change in one's mindset. At the onset of this project, I somehow wanted to change my mindset. I felt something missing, I felt the grips of apathy, banality and desecration. But how do you change your mind; the mind changing the mind?

I know that my own mindset is almost invisible to me. It is the way I see the world, the lens I see the world through. It is like the grammar of our mother tongue – we would never know it to be there unless it had been pointed out to us. Therefore the inventions of words that point to this invisible concept can help us grasp what we are looking for. Under my way to a new mindset, I have come across different metaphors for the elusive concept of 'mindset'. These alternate concepts provide me with a broader understanding of the 'holy grail' of this search.

**Imagined order:** Reminding me that regardless of which mindset I see the world through it is an order I use as an interpreting interface between me and the world and that this order is imagined. The order or patter I see the world with is just one of many possible patterns. It is also 'imagined' in the sense that I can change the order if I can imagine something else that is sufficiently consistent with what I experience.

**Operating system:** This metaphor makes it clear that the mindset establishes a system for how one operates in the world. And just like in the world of computers and electronic devices, the operating system determines what programs or apps can be run. At the same time, the metaphor indicates that though the mindset has an element of 'imagination', there are also non-negotiable basic specifications that need to be in place for an operating system to work at all.

**Heart set:** The mindset is not only located in the mind but is also present in how we perceive the world wordlessly through the heart, our body and nervous system. This is an interesting alternative concept for 'mindset' as it centres the attention in the heart or the body rather than in the brains or the head. Just simply shifting the attention to the heart changes the world.

**Decision pattern:** From the outside, from other peoples' perspective, our mindset is most easily discernible by the pattern of decisions we leave behind us. An outsider can only perceive what we choose to say and do, not what we think and feel. We cannot make better decisions than what our mindset allows us to. The ability to create integrated decisions that include both the real needs and meets the complexity of reality just right is based on what we can perceive. Some words or ideas we cannot hear, before we get an inner concept of them; until then the picture people may try to draw appears to be meaningless.

**Worldview:** The mindset under development is not just a mindset about water, rather our mindset about water is a continuation of how we view our whole world. Also here things are well connected.

What we want to do is not change our mindset to something specific, like a kind of brainwash – that is not the aim. A better metaphor for what we are trying to do is to let the heart unfold naturally. Others use the concept of getting increasingly enlightened. Regardless, it is not about pushing a certain doctrine into the brain. But rather to find one's own meaningful way forward. In another sense, the process of 'heart development' is different from the process of 'mind development'. Mind development is about understanding what is outside us. Heart development is about understanding what is inside us – or perhaps not understanding, but rather listening to. 'Heart development' therefore, more than anything else, is an exercise in listening.

Try to hear the calling of the heart towards health and sustainability. Try to dig deeper in you to get an inner understanding of what sustainability means in your own system. The world is full of audacious but superficial speeches about sustainability. But 'talking and talking' will not make the change happen. Understanding the concept of sustainability is a personal challenge. It has to start in a new personal and collective practice of the actual activity of 'sustaining'. But what is it that we want to sustain? What joy does it give us? When do we hear and feel that joy in us?

The heart can only be felt individually. It will not be enough to hear or read about the heart. One must find one's own practice. Over time, I have read many accounts about the heart, yet I believe that my own experience and 'explorations' are much more vibrant than any second-hand account – to me. It is not possible to rely on someone else's experience or even 'objective rational research' on the matter. It may, in fact, be our tendency to try to see ourselves from the outside – to try to apply a scientific principle to ourselves or our heart that is a part of the malfunctioning. It is the wrong perspective to take. Exploring the world of the heart is 'an inside job'. It is possible to learn where and how to look by listening to other's accounts. But the first-hand experience is essential to become solidly entrenched in the 'value system of the heart'.

The importance of a strong personal entrenchment is best understood when the difference between our potential imagined and current impact is clarified. If our internal compass on where and how we want to move forward is unclear, we are bound to get lost. And if we by luck or hard work happen to move into a position of power where we may actually have a real chance to change something, we may quickly come under pressure and revert to our old 'well known and safe-feeling mindset and habits'. It is quite demanding to bring your personal vision all the way through this. There is pressure from superiors, employees and peers, there are time pressure, there is pressure due to lack of domain knowledge. With only a short time to choose the road forward, we may soon get lost if not solidly entrenched. Hence the work of getting entrenched in a new mindset is a long-term commitment, where we can only change little by little approaching a new ideal. However, if we do not do this internal work intelligently, disciplined and consciously, the old mindset and the old habits will move us like an unstoppable freight train.

At the beginning of this 'project', I made a vow, a solemn promise to myself that consisted of a commitment to the heart. In the beginning, I hardly knew what that meant, and I felt it was a difficult choice, a choice with many internal conflicts. I was conflicted because on the one hand I had this clear notion that this would be of almost life-or-death importance to my life, but on the other hand I was afraid what it would mean for my identity. Would it be a loss of all thinking, reasoning and science – unto which I had put a lot of my identity? Would it mean doing reckless things on the spur of any emotional impetus? Would I see things about myself hidden in the shades that I didn't want to see? Would I have to make drastic, irreversible changes to my life? And how would all this influence my family, friends, work, colleagues?

At the same time, there was this strong calling for it. As if I would regret it on my death bed if I did not follow through. Additionally, there was a pinch of curiosity and adventure. What would this different starting point mean?

The fear was real, and it felt like I was going under some kind of imaginary water line and I almost literally feared that I would drown. Looking back, I didn't drown but came back to the world with essential insights for refocusing my future work. I wanted to continue using these new eyes with their new poetic perceiving – to be able to see from the heart in mildness. I wanted to enhance further my ability to feel life – the process of it, the tic-toc of it, the miracle of it. I wanted to protect and nurture the heart as a fragile inner sanctuary. I had glimpses of emotions of suddenly 'standing up in full height', which felt like something important 'clicked in place' and changed my whole posture. These different inner experiences of how to be in this world changed my understanding of my water vocation; it changed my view on water and changed my actions. These findings are of course my findings, but at the same time, they seem to have a universality.

This 'heart-mode' is not consistently present in my life, but now I recognise it, and I can find my way back to it. In the beginning, I found it difficult and even fearful to embrace the heart entirely. I still find it difficult, but the fear has gone. This is not 'practice makes master' in the usual sense. It is not like learning math or any kind of science. It is clear that my 'heart capabilities' increase, but it is more as if I get access to more and more within me. Whatever the wisdom might be, it was already there – just out of reach or incomprehensible. And as I walk further into it, I find again and again that there are more levels and nuances to it and that it takes time, learning. It takes time to remember to focus, to keep focus and overcoming the lack of patience with learning it.

*'It is only in the heart that one can see rightly; what is essential is invisible to the eye.'*
Antoine de Saint-Exupéry (1943)

> **Your reflections:** How is your sense of your own physical heart? Do you know where it is located precisely? Do you know how it works? Do you know the shape of it? What happens if you attempt to connect your thoughts to your heart? Do you have any kind of 'spiritual' thinking or feeling about your heart? What does your heart mean to you? If you look at your purpose in life as you formulate it to yourself, how does the purpose you speak of relate to the purpose you originally thought you had? How does it relate to the purpose as it would be perceived from the outside? How does it relate to the most potent purpose you could formulate?

## A SCIENTIFIC APPROACH TO THE HEART

It was not that I was unaware of the position of the heart in literature and mythology. However, what surprised me, when looking into the more scientific side of it, was that some of the more poetic ideas about the heart were actually 'retrievable' in measurements. This insight made the task of connecting to the heart easier because it made logical sense. But at the same time, it also made it more obligating because with it came an understanding that the idea of the heart was not just a metaphor, but something much more direct and practical. It had real practical meaning for me and those around me. Suddenly, a day with low heart connection felt less worthwhile than a day with a strong heart connection. For a while I felt I was not 'working hard enough' if I did not remember heart focus. But again, that is not how the heart works.

Regardless, the research at HeartMath Institute (Childre *et al.*, 1999) shows that there are several communication lines between the heart and the brain. It seems however that the communication from the heart to the brain is more information-rich than the connection the other way. The signals from the heart to the brain include: (1) neural via the vagus nerve that goes up the spine and enters the brain in the medulla, (2) biochemical via various hormone systems, (3) biophysical via changes in blood pressure and (4) electromagnetic (the heart rate can easily be picked up by measuring the electromagnetic field generated up to 1 meter away from the heart). While there is also information going from the brain to the heart, it was found that the heart was autonomously in control of which signals to react to in contrast to other more obliging bodily functions.

Secondly, by measuring heart rate variability, i.e. the change of the rate of the beating of the heart, it is possible to observe how well a person is balanced. Self-induced meditative states can also be observed as a clear coherent signal from the heart. The heart rate has been shown to vary from heartbeat to heartbeat. During good meditative practices, the heart rate varies harmoniously like a curve (it is not the ECG that varies as a sinus, but the distance between peaks).

Thirdly, the signal from one person's heart can be picked up by another persons' mind and heart. If one person touches another the signal from the person's cardiac electrical signal (ECG) can be observed in the other person's brain waves measured by EEG. And it is possible to reach a state of entrainment of the heart beatings in a group working well together. Hence one person's heartbeats can unknowingly affect those around him. It appears that heart signals are being broadcasted and received invisibly and unknown between people all the time.

> *'Early in the Institute's research, we observed that when negative emotions threw the nervous system out of balance, they created heart rhythms that appeared jagged and disordered. It was easy to see that a chronic state of nervous system and cardiovascular imbalance would put stress on the heart and other organs that could potentially lead to serious health problems.*
>
> *Positive emotions, by contrast, were found to increase order and balance in the nervous system and produce smooth, harmonious heart rhythms. But these harmonious and coherent rhythms did more than reduce stress; they actually enhanced people's ability to clearly perceive the world around them.'*
>
> Doc Childre *et al.* (1999)

When I learned of the research of HeartMath Institute, I got quite interested in measuring the effects of my meditative practice. I found the idea of this type of bio-feedback intriguing and was curious to measure the impacts my meditation had on my heart rate variability. So, I acquired the measuring device from HeartMath Institute to make my own personal experience.

The device is simple and works by measuring the heartbeats through an ear clip measuring the pulse of the blood flow in the ear lobe. The core of the technology is analysing this signal to measure the coherence, i.e. the harmoniousness of the change in heart rate over time. The ideal pattern being close to a sinusoidal wave of continuously increasing and decreasing heart rate. In the beginning I 'brain-talked' a lot about the exact algorithms for this measurement. Soon, however, I found out I had to let go of these 'brain talk' discussions, as they invariably and ironically led to a decrease in heart coherence.

The short instruction of use basically comes down to: breathe as if through the heart, nurture positive and calm emotions and stay alert.

The device is followed by an app that tracks 'performance' and that allows the user to grow better over time with a simple 1–4 star difficulty setting. The setting regulates whether a given level of coherence (yes it boils down to one number) is characterised as good (green), reasonable (blue) or not so good (red). This feedback is given regularly both as colour and sound every few seconds.

At first, I was frustrated with the system because it was an outside 'judge' of an inner experience. My practice had the exact opposite objective, to emphasise the inner experience over the outer experience. And suddenly this outside-device was judging my 'performance' from the outside. But gradually I came to terms with this and found a way of integrating the sound as 'a sound of my heart response'.

As most people who meditate, I often experience after a while that thoughts start wandering. The interesting thing was to observe the heart feedback system responding to the nature of my thoughts wandering. It turned out that the heart rate was susceptible to the subtle changes in the nature of my thoughts. It picks up on things that were slightly different from what I expected. But over time, it came to make sense and hence – I am sure – changed my perspective and weeded out some of my less healthy thinking patterns.

Mostly, at the beginning of the meditation, it was relatively easy to gain a high level of coherence, but after a time, the heart rate variability would drop. Often the 'thing' that interfered could be categorised as 'egoic thoughts' – the not so pleasant thoughts. It could be I was just a bit too proud about the result. Or it could be the opposite when I saw myself as a victim of events that annoyed me. A recurring problem would be to achieve high coherence quickly and then being just a bit too proud about that. This would invariably and ironically lead the coherence to plummet within seconds. These short-term 'internal calibrations' were very informative as to how little it takes to get out of balance and how fast the heart – and ultimately the body – responds to that. Doc Childre notes that: 'If we live in a stressed-out state all the time, we become used to imbalance'. It made me more cautious about my thoughts and speech in the following time. It became apparent to me how often I casually think, rage and joke about this or that – and how much that costs every day in lack of coherence. I would not have noticed that or known it before the means to measure these effects became available to me. Since I could feel a long term (days/weeks) improvement in my inner balance based on these experiments, it was clear that coherence was indeed very beneficial for my exploration and for me personally.

However, I also experienced that sometimes it was small adjustments that made the coherence soar. Sometimes, it felt impossible to 'make the heart come into coherence'. But then at some point, I would give up and look into the blue sky above me – and that would cause the heart to jump into coherence. So even if the use of the mind and the thoughts that go through it are essential in this kind of 'heart training', the heart cannot 'be willed' into coherence. It seems that it requires the mind's cooperation, but the mind is not in *control* of coherence beyond setting the minimum circumstances for it to happen. These circumstances are about mildness, gratitude, alertness and truthfulness.

I also found that regularly getting into coherence changed my day-to-day emotional landscape into a higher degree of integrity, self-command and calm. I found myself able to better and more easily handle difficult events that would earlier have thrown me into frustration, anger or confusion. Before the experiments with the method, I had the idea (or hope) that I would be calm and un-stressed all the time. However, to my surprise, I found that coherence was not only about relaxation. Some of my most stressful days with a lot at stake could result in quite high coherent results. The highest level of coherence I have measured yet was actually on such a day, where everything 'seemed at stake' and I had to be 'high-performing' to make things fall into their right place. I believe that what helped the coherence stay high was that I felt I was capable and reasonably 'in control'.

Another significant effect of the long-term use of the device was that I became more grounded. While earlier after a hectic day I could be doubtful in the evening, questioning my moral ground for this or that decision or this or that approach to another person. I could wonder should I have been kinder, should I have been less aggressive, or should I have held my ground more firmly etc. The continuous feeling of calm and groundedness meant that I felt much more at peace with my actions. There was a kind of predetermined forgiveness about it all. That regardless of what effects my actions would result in, I knew I was calm and balanced at the time of the action, so it was by definition the best action I could come up with. Had I done it again my actions would have been more or less the same. Hence, I agree with the inventor Doc Childre, that 'the heart isn't mushy or sentimental'. It is not about 'rainbows and unicorns'. It is about kindness, integrity and a grounded balance.

The research and experiments at HeartMath Institute gave me an understanding from the outside in. It set a context; supporting a hypothesis of the meaningfulness of the heart commitment.

One could see heart intelligence as a different intelligence than traditional intelligence. It requires the cooperation of the brain. The brain needs to surrender its stories and the protection that the stories provide. It is in a sense the eternal battle with the ego. But it does not take the form of a religious, hard morality and discipline struggle. Rather it is a 'game of surrendering' in terms of reckoning that there is something silent that is grander and that I/you/we need to place ourselves, our attention in that grander silence. Putting oneself in service is not a one-time event like signing in, but rather a continually repeated process of committing full-heartedly and humbly.



## EMOTOS EXPERIMENTS

When I share with others what I am doing in this domain of heart and water, many ask me: so what do you think about the Emoto experiments?

Masaru Emoto (1943–2014) carried out some curious experiments linking water and the heart and emotions through the observation of crystals. I admit, I still don't know what I think – if I believe it or not; if I think it is science or not. But it is extra-ordinarily interesting that a person applied this focus throughout his life for these experiments. And the results he claims to find are … beautiful.

Ice crystals are formed when water vapour freezes into ice in the sky. The crystal starts as a hexagonal prism from where the crystal branches out in six dendrites, which may branch out further in a fern-like structure governed by the temperature, pressure and water-vapour content in the air the crystal travels through on its way to earth. This results in trillions and trillions of ice crystals landing on earth annually. Still, it is claimed that no two ice crystals are the same. Whether this statement is accurate is unprovable; however just the fact that it is probably true is impressive – and worth an admiration similar to having 'stars falling on your coat' as Thoreau suggests.

*'How full of the creative genius is the air in which these are created. I should hardly admire more if real stars fell and lodged on my coat.'*

Henry David Thoreau (1856)

Japanese Masaru Emoto was a man of science. But above his scientific studies, he wondered why people were unhappy. He thought to himself: who would be able to reply with a resounding 'yes' if I were to ask them: 'Do you have a sense of peace in your heart, a feeling of security about your future, and a feeling of anticipation when you wake up in the morning?' One day he read – as above – that no crystals are alike. He had heard this already as a child, but it collided at that very moment with another idea on his mind. The idea that water, in some mysterious way, sees and listens to its surroundings and stores the impression of this in its molecular structure – a controversial idea that in respected scientific circles was (and is) considered to be complete and utter nonsense.

However, regardless of this scepticism, he thought that perhaps the crystal-making process is a way to see with his own eyes what he suspected to be invisibly true. Perhaps the structure of the ice crystals would depend on the emotions that surrounded their making?

Emoto wrote about his experiments with ice crystals throughout his life, most famously in 2004 in the book 'Hidden messages in water' (Emoto, 2004). Masaru Emoto described himself as an 'original thinker using scientific methods'. In the story of the idea, he tells that it took his research assistant months even to figure out how to form crystals of water in the laboratory – trying different ways to maintain the temperature conditions for the formation of crystals artificially.

Emoto's experiments with the formation of water crystals showed him that conscious thought, positive or negative, influences the degree of beauty of the crystals formed. He found that beautiful crystals were formed with conscious thinking and feelings of gratitude, kindness, courage and peace while the crystals subjected to feelings of hate, anger; fear and anxiety were less sophisticated, less refined and asymmetrical – ugly, compared to the other crystals.

While this seems like a highly improbable hypothesis seen from a scientific mindset, it stroked an intuitive cord with a lot of people and the book has been read by thousands of people worldwide – it became a New York Times bestseller. If we entertain the thought of 'what if…', we are led down new alleys. What if water responds to how we handle it, think of it and speak to it? How would the water we drink and consume reflect this – and what would that mean to our well-being? What would it mean if the water in our bodies were sad and tortured water? What would make water happy?

*'The earth is searching. It wants to be beautiful. It wants to be the most beautiful that it can be.'*

Masaru Emoto (2004)

I more than hesitate in accepting the basic premise of the experiments and for many years people have led my attention to various strange water theories. I have generally dismissed them quickly and arrogantly. But there is something about this idea that intrigues me – true or false. Emoto's idea and findings are that the emotional vibrations that water is subjected to affect the water and can be seen in the beauty of the crystal. According to Emoto, even relatively subtle changes to the application of words can be detected. The inviting words 'Let's do it!' creates a harmonious crystal, while the commanding 'Do it!' does not. Similarly, the crystal forming is affected by music played to the water. And finally, water is affected by its treatment, systematically chlorinated water from water utilities, as in for example Tokyo, is hardly able to form crystals. Lake-water blessed by a monk showed to be more beautiful than the water was before the blessing. Essentially, Emoto's point is that water is a 'sensitive mirror' to the world surrounding it – like a kind of a 'master listener'.

Emoto believed that the better the vibrations are in harmony with nature and earth, the 'happier' the water is. He knew that this would be difficult to believe. He described this problem as:

*'Traditionally speaking, anyone who says that consciousness has an effect on the physical world risks certain ostracizing for being unscientific. However, science has progressed to a point where the failure to understand consciousness and the mind limits our understanding of much of the world around us.'*

Masaru Emoto (2004)

So, if we play along with this and imagine that the crystal forming capacity of water forms a communication or recording bridge to a kind of extra-ordinary world, what wisdom does Emoto derive from his life-long work with this?

As I would have done, had it been my scientific experiment, he experimented to find words that expresses themselves in water with the greatest elegance, intricacy and harmony. Unsurprisingly (when accepting the premise) the word 'love' scores very high. However, he finds that 'gratitude' and 'love' together gives the best result – and the crystals formed seemed actually to be closer to 'gratitude' alone than to 'love' alone.

*'Love tends to be a more active energy, the act of giving oneself unconditionally. By contrast, gratitude is a more passive energy, a feeling that results from having been given something – knowing that you have been given the gift of life and reaching out to receive it joyously with both hands.'*

Masaru Emoto (2004)

The final point Emoto makes based on these observations is that we should give water respect. Water makes up most of the weight of a human being. Life is so deeply linked with water. Emoto points to a kind of pollution different from traditional water pollution; a metaphysical pollution made up of negative thoughts and emotions – or even worse as Emoto found: the lack of care and attention. This is seen in current days lack of reverence for water and the treatment of water as a mere commodity. The quality of care is in profound contrast to the respect paid towards water in earlier times:

*'The important thing is that we recover our desire to treat water with respect. In our modern culture, we have lost our attitude of respect for water. In ancient Greece, people paid true respect to water, and many Greek myths are based on the protection of water. But then science appeared, and rejected these myths because they were not scientific.'*

Masaru Emoto (2004)

Emoto is a man of science himself, so it is not science per se that he rejects. What we must remember is that science is only a method to learn about the world – a method that moves us inch by inch closer to better models of the worlds. Models that are useful, not necessarily profoundly true and not as a philosophy of life that can stand on its own. Science is a tool of understanding. Science is also a key tool when it comes to understanding the sustainability crisis. A crisis that would have been difficult to understand the scope of without science. It is what we do with the science that we should be aware of. Using science to reject myth is a misunderstanding of both myth and science.

*'Perhaps we are finally beginning to see that the direction we are moving in leads nowhere. We have sacrificed too much in order to secure the riches of life. Forests have been destroyed and clean water lost, and we have cut up and sold the earth itself.'*

Masaru Emoto (2004)

The road forward is to use science from a different starting point and with a different aim. To ground ourselves in making life possible, relax our fear and grip for control and instead keep a deliberate focus on celebrating life and its diverse richness. This approach has ample space for facts, stories and myth.

*'We need to feel gratitude for having been born on a planet so rich in nature, and gratitude for the water that makes our life possible.'*

Masaru Emoto (2004)

Some attempts have been made by Emoto and people around him to work this out scientifically – to prove the hypothesis employing blind tests. But the difficulty of the story is the same as always when we work with consciousness. Our minds so wish for a good story that it will easily make one up where there is a lack. Proving that the mind did not make this up somehow is difficult – especially in the sense that what is attempted to be established is that the mind makes up, or at least influences, reality.

I am however, struck by Emoto's attempt to find proof and use science to prove a connection between consciousness and water. Our civilisation has invested so heavily in science and the scientific method. I wonder if 100 years from now they will think of us as stubbornly stupid for clinging to a hope of 'science to explain all' or on the contrary that this epoch was the last rattling of the belief in something beyond the rational.

Some say his interpretation is not valid and cannot be replicated. But perhaps again the scientific truth is one thing, and the poetic side of the systems understanding is something else. The work of Emoto became very popular among people who sought for meaning and sensed a kind of truth that these results articulated. At the same time, if we take this interpretation to our heart, something new happens in how we understand water and how we would treat it. The history of the water we drink becomes important in a new way that the current scientific sustainability story cannot do on its own. It provides a different impetus for care and personal starting point.

Looking at water through Emoto's lens gives a gentler touch – a more sacred outlook on the work with water. Having such a story as a lighthouse in the back of my mind changes my approach compared to a clean, rational approach. While this may not be scientifically valid, neither does it mark a return to olden day's superstition, with all its violent shadow sides. I think that a combination of interpretations that beautifies our world with a rational, scientific approach may work as a functional integration in the new mindset. When a movement of 'naturalness' argues that we ought to drink untreated water and then people get ill, that is not a functional combination. But if I apply my science and have a sacred interpretation of water at the same time, I can move into a deeper understanding from where better questions arise.

So, in essence, the opposite of banal thinking is to apply higher, deeper and broader thinking as well as looking into the shadow. Thinking higher means having higher aspirations for our work. Thinking deeper in this context means understanding ourselves better and using our emotions as refined navigation tools. This way of systemic understanding is relevant for many different types of systems: ourselves as a system, the system of our utilities and the global system. It can also be used to understand roles, i.e. the role organisations play in the global system or the role a person plays in a given organisation. So the point is not to act at the spur of emotion, but to think deeply about our feelings to understand the systemic information they carry in their reaction. Thinking broader means broadening our scope of interaction, i.e. broadening our personal and organisational interaction span to more stakeholders. It also means taking more aspects into consideration and looking at situations from a higher-level helicopter view. Finally, we have to turn around and look at our shadow and into our shadow. We cannot allow ourselves to be too ignorant about the shadow effects of our action, inaction and ignorance – we need to understand and handle our levels of inertia and ignorance. So, this is a call for feedback from people around us and being attentive to this feedback so that we realise in good time when we are out of our competence comfort zone. When that happens we might overlook issues that people better versed can see. To be helpful is only half the equation, it is just as important to ask for help genuinely and authentically to reduce the shadow effects of our work.

> **Your reflections:** What do you think of Emotos experiments? Are you curious? Do you feel a softening? Do you reject it as impossible – if so how fast did you make up your mind and on what grounds? Fraud? Ambiguous?

## THE ROLE OF THE HEART IN THE LOCAL PLACE

To attempt to transform this inner wholeheartedness to something that is visible or can be felt from the outside, we need to connect to the outside – which can be done in different ways.

Norwegian professor (philosopher and keen mountaineer) Arne Naess offered a concept called 'deep ecology' (Naess, 2008). 'Deep ecology' stands in contrast to the classical anthropocentric environmentalism's 'shallow ecology', where 'the environment' is something just surrounding humans. In anthropocentric environmentalism, the environment is wittingly or unwittingly reduced to a resource for human societies. We speak of water as a resource or we use it as a dumping ground/basin for waste and wastewater. I find the novel concept of 'ecosystem services' as having the same ring to it. The argumentation in 'shallow ecology' is that we need to preserve nature primarily for our own survival, i.e. because it provides us with resources. 'Deep ecology' on the contrary is about understanding the local ecology deeply and appreciating the natural world as a habitat that we share with our fellow animal. In this view, nature is not valued according to the instrumental value for humans but has its right to exist.

Arne Naess was a professor at Oslo University from 1939 to 1970. Arne lived a significant part of his life in a place called Tvergastein in the mountains; a place in the arctic, where he built a small house to stay in and to study the ecological habitat around him. This place was an essential inspiration to his work, a place where he investigated his sentiments towards 'place'. By spending days and months studying the area he developed a personal bond with the place – and this bond is at the core of his basic idea. The bond defined what he called 'the ecological self'.

Arne Naess suggests a deeper understanding of the place one lives on/in. In that way, a place becomes a Place and a home becomes a Home. Arne Naess suggests two ways of understanding ecology and introduced the notion of deep and shallow ecology.

> *'When the majority of people were living off the land, with little mobility, it was natural to feel at home at certain places. One stayed at home, left home, or went home. But home was not a building. The advertising of homes to be bought is not an offer of a home in the connotation relevant in our analysis. Home was where one belonged.'*
>
> Arne Naess (2008)

Home in the sense that Arne Naess talks about it delimits an ecological self, where a person – the I – is deeply integrated a 'home ecology'. Having moved around a lot in my life, I can feel this sense of 'home' when I enter the town of my childhood. I register that I breathe differently there. Arne Naess contrasts this sense of Home with the word 'environment'. A word that is much more circumstantial and without thousands of the hundreds of relationships of belonging.

> *'But humanity today suffers from a place-corrosive process.'*
>
> Arne Naess (2008)

Arne Naess moves the analogy further than the sense of childhood home. As an example of what it means to identify he describes a peculiar case of his identification with a flea:

*'I was looking through an old-fashioned microscope at the dramatic meeting of two drops of different chemicals. At that moment, a flea jumped from a lemming that was strolling along the table. The insect landed in the middle of the acid chemicals. To save it was impossible.'*

Arne Naess (2008)

When he sees the jerking movements of the flea, he feels pain himself. He registers both what is happening to the flea and what is happening to himself. He calls the feeling 'a painful sense of compassion and empathy'. He interprets this inner sense as identification.

Is that the process of identification? To see oneself in the other? To register, to understand, to sense, to intuit the same process of living and dying. To sense that more than anything else, one is that process. That process is what makes us the same from birth through adulthood and all the way to death. That process is recognisable in the ecology around us – if we take the time to know it and open the sensitivity to see it.

*'If I had been alienated from the flea, not seeing intuitively anything even resembling myself, the death struggle would have left me feeling indifferent. So there must be identification for there to be compassion and, among humans, solidarity.'*

Arne Naess (2008)

The peculiarity of the example is to show that there isn't a limit of insignificance to the feeling of the ecological self, to the identification with even small insects, that we are able to understand it's feelings. That he holds these feelings also shows that this empathy with every living being – even a flea – is not something we need to learn. It is something that is already there. And we get access to it when we pay a certain kind of attention. On the other hand, we as humans can cut ourselves off from all these empathic feelings if necessary, and our culture in many ways supports this cutting off. So, to nourish these feelings and this attentiveness is already a radical change. If this example seems too bizarre, consider this other example.

*'Tragic cases can be seen in other parts of the Arctic. We all regret the fate of the Inuit, their difficulties in finding a new identity, a new social self, and a new, more comprehensive ecological self.'*

Arne Naess (2008)

Arne Naess recounts a case in Arctic Norway, where Lapps had been forced to move due to a diversion of a river for hydroelectricity. Similar cases are seen in many places where indigenous people are forced to change their homeland and traditions.

*'In court, accused of an illegal demonstration at the river, one Lapp said that the part of the river in question was 'part of himself'. This kind of spontaneous answer is not uncommon among people. They have not heard about the philosophy of the wider and deeper self, but they talk spontaneously as if they had.'*

Arne Naess (2008)

Hence, the idea of deep ecology is that we as persons can develop our sense of our ecological self – or perhaps reconnect with it as it lies dormant in us. When we become aware of this larger ecological self, we become more than our usual day-to-day small ego. This understanding is closely connected to our understanding of our spirit. Hence this is one pathway to the spiritual or religious for many people. Surprisingly, though religions may often divide, in terms of deep ecology, various religious beliefs may inspire the same kind of insight and compassion for all living beings.

*'One must avoid looking for one definite philosophy or religious view among the supporters of the deep ecology movement. There is a rich manifold of fundamental views compatible with the deep ecology platform. And without this, the movement would lose its transcultural character.'*

Arne Naess (2008)

People who recognise this feeling of the ecological self may reach a point where the act of nature preservation or environmental concern and work is not only an obligation or a duty but instead becomes what Spinoza calls a 'beautiful act', in contrast to a 'moral act'. A moral act is informed by a sense of duty and carried out in a disciplined way and despite self-interest. The same act carried out as a beautiful act is carried out in alignment with one's self-interest. The act is recognised by the ecological self as an act for the self also.

*'Now, my point is that in environmental affairs, perhaps we should try primarily to influence people toward beautiful acts. Work on their inclinations rather than morals.'*

Arne Naess (2008)

I have experimented with this concept. It is much more fragile than morals, discipline and strictness. But the feeling of this approach is like an opening and a deepening. There is a sense of connection and beauty. To be able to keep in touch with this fragile emotion still requires a good portion of discipline – especially in our current culture and especially because the effects of this openness is slowly accumulative and hence difficult to appreciate or even to discern in the beginning. It requires one to be still to be able to sense.

*'As I see it, we need the immense variety of sources of joy opened through increased sensitivity toward the richness and diversity of life and the landscapes of free nature.'*

Arne Naess (2008)

Deep ecology points at something fundamental for the notion of water stewardship while it has almost nothing in common with the current concept of 'water and wastewater system operator'.

Naess describes the two different types of ecology as a shallow ecology where human beings are on the top and outside of nature and deep ecology where humans see themselves as part of the whole ecology. This captures a key point in the heart approach. The heart approach is a way of becoming part of it all rather than standing outside or above. It is an integration effort.

Naess sees the ecological movement as the third great movement after the social movement concerned with equality for all and the peace movement. In this evolvement of engagement, there is a sense of caring in eccentrically larger circles around oneself. This describes a journey starting with self-interest and expanding to include family, to a wider and wider part of one's community at some point also including people of a different race, religion etc. At some point this could expand to include animals and eventually all living beings.

The self increasingly comes to an understanding of itself as first reliant then dependant on and finally as part of a greater and greater sphere. With this worldview, the care for nature is not a weight of responsibility or moral obligation for external entities but rather an act of self-care. Everything becomes a 'beautiful act'.

For most people this is not such a long stretch. In E. O. Wilsons essay Biophilia, he points out to us our feeling of 'love for nature'. This feeling of 'biophilia' can be recognised as our attraction to baby animals, love and care for our pets and the way we surround ourselves with flowers. There is this natural and innate love and care for natural things. It is not difficult to imagine how we could develop this sense further and thereby generate both internal gratification for our benevolent care and protection for nature – not requiring some self-sacrificial burden.

To make a place to a Place in practice, Naess suggests a practice of extensive study of the place where one lives. It is as if through his understanding of Tvergaisten and the full potential of all its constituents he gets to understand the Place and it becomes his Home. He concludes that his role in the place is to observe and understand. It is not to actively govern or even facilitate. By identifying with this role, the Place becomes – in a sense – a part of his deeper ecological self.

When our understanding of our place deepens and widens, we will gain a more comprehensive understanding of the role of water. And that may change our mind about water as a resource primarily for our direct needs and purposes. What we will naturally realise is that water has many other purposes than providing water to the city, purposes that when unmet means actual losses; loss of life and vibrancy in the place we live in. We must carefully weigh what is acceptable and what is lost when we take water for our society, and perhaps how we can deliver the water back. However, if our decision of 'our needs first' is made again and again everywhere, the poverty of our place becomes so deep that in the end, we cannot even remember the natural richness of our place.

> **Your reflections:** Is your ecological knowledge deep or shallow? What are your sentiments to being in nature? How much of your days have you spent in nature? What are your childhood memories of the forest, the beach, the meadow, the desert – or whatever nature you were most in? If you were to do one thing to deepen your ecological connection, what would it be?

## A PLACE UNTOUCHED

When I was a child and fairy tales were read to me, I was always impatient with the lengthy descriptions of nature. On and on it went with the different trees, the birds, the insects, the sun, the stream, the air, the sounds, the smells. All this description seemed very circumstantial to the real story, the true adventure. On occasions, I tried to tune out the descriptions, but I couldn't, so eventually, I surrendered and instead listened to the long descriptions as a secret 'training of my patience'. I tried to reconstruct the exact natural place in my mind's eye.

Despite my belittling of the descriptions of nature, they must still have made an impression on me. Because due to these mental pictures of the forests, I have often been somewhat disappointed with the real experience of the woods I have visited. They often appeared to me to be considerably less magical than the wordy descriptions I had heard from hundred-year-old fairy tales like the ones by Hans-Christian Andersen. I have long ascribed this discrepancy to an over-active poetic imagination in the face of reality or the writers of fairy tales being poetically exaggerating or painting pictures of forests more magically than reality for the reasons of emotional effect. But then I learned about Białowiea Puszcza and I really got a Eureka-moment about what has happened.

Białowiea Puszcza is a forest primaeval, the last(!) remaining of temperate Europe. It is the approximate size of greater Paris ($2100 \text{ km}^2$) and lies on the border between Poland and Belarus. The forest is not wholly untouched, there are signs of people having lived in the forest through the ages, but they are only signs. The forest has not been transforming due to exploitations as in other wood areas.

The history behind this special reserve involves several special events in the history of Europe, dating back to the 14th century where the Lithuanian Duke Władysław Jagiełło pronounced the area a royal hunting reserve. During a period where it belonged to the Russian empire, it was protected as a private domain of the tsars. There were threatening years around World War I and World War II, between which the forest was declared a Polish national park. During World War II, Soviets and Germans were taking timber from the forest, but Hermann Göring saved the forest by declaring it completely off-limits for everybody except for those he would allow in there. When Stalin took over the dominion, he allowed

Poland to keep two-fifths of the forest, and it has succeeded in not being seriously exploited up until today, where it is protected as a UNESCO World Heritage Site.

I have not visited the forest myself. But I have visited it in my imagination by reading about it, watching videos from it, and tracking it all round on Google Maps. So, here in this book, I will allow – maybe to test the reader's patience – a description of Białowiea Puszcza. Not as I see it, but as writer Alan Weisman (2007) sees it.

> *'Here, ash and linden trees tower nearly 150 feet, their huge canopies shading a moist, tangled understory of hornbeams, ferns, swamp alders and crockery-sized fungi. Oaks, shrouded with half a millennium of moss, grow so immense here that great spotted woodpeckers store spruce cones in their three-inch-deep bark furrows.'*
>
> Alan Weisman (2007)

This sense of ecological diversity is fused by the sound and sense of being in the forest:

> *'The air, thick and cool, is draped with silence that parts briefly for a nutcracker's croak, a pygmy owl's low whistle, or a wolf's wail, then returns to stillness.'*
>
> Alan Weisman (2007)

What I did not know or understand before, but what is becoming more and more apparent, is that a 'magical' process happens over time as nature is allowed to emerge on its own. The process of death and decay is an important ecological component.

> *'Almost a quarter of the organic mass above ground is in assorted stages of decay – more than 50 cubic yards of decomposing trunks and fallen branches on every acre, nourishing thousands of species of mushrooms, lichens, bark beetles, grubs, and microbes that are missing from the orderly, managed woodlands that pass as forests elsewhere.'*
>
> Alan Weisman (2007)

And this non-utilitarian approach to the forest has made room for animals and species that can't live without this decaying diversity and hence is not found in ecological biotas that has not been left alone for this long. The forest manages itself and creates niches. Białowiea Puszcza is the only place with nine European woodpecker species, because some of them only nest in hollow dying trees.

> *'Together those species stock a sylvan larder that provides for weasels, pine martens, raccoons, badgers, otters, fox, lynx, wolves, roe deer, elk, and eagles. More kinds of life are found here than anywhere else on the continent [...] The Białowiea Puszcza is simply a relic of what once stretched east to Siberia and west to Ireland.'*
>
> Alan Weisman (2007)

When I listened to this description of the forest and compared it to my local experience of parks and forests, I find the same discrepancy – making me want to read all the fairy tales of childhood again; listening again to understand what could have been there, in my local forest. I want to understand this feeling of unknown or unrealised deprivation – a deprivation of something I have never experienced. A deprivation that has come into our culture so slowly that it was only noticeable from generation to generation by the most apt observers. A slow taking out of riches from nature. Recently, I heard a researcher describe many Danish forests as no more than 'fields of trees'.

Alan Weisman tells about the difficulty protecting even this small area in Poland and Belarus today. Forest experts are so compelled to go in and 'make their trade'. It is as if they can hear the calling of all the accumulated value of natural material. It is so tempting to go in and convert the forest goods to monetary value.

A movement for 'recreating' original forest is forming these days. What I find fascinating in this endeavour is that the time horizon to re-create something like the forest on the border between Poland and Belarus is 500 years! Hence a protection project should ensure that the forests are protected, at least for the next 500 years to succeed. Maybe I could pass on the baton to my children, who could pass it on to theirs. In that case, that chain would have to work for something like 20 generations. To have such a forest today would have required protection since the end of the middle age and the beginning of the modern ages, i.e. since the time of Hernán Cortez, Christopher Columbus and Vasco da Gama. Through all the tumult of contemporary history, it would have to be protected.

> **Your reflections:** Nature around you is possibly considerably less rich than it could be and less than you imagined. How does that feel?

## THE ROLE OF HEART IN RELATION TO THE EARTH

After having looked into whole-heartedness in a personal and a local perspective, it is now time to investigate what that means on a global scale. Again different words create different associations. If we speak of 'globalisation', we have one kind of thinking. If we instead use the word 'spaceship Earth' as Buckminster Fuller did, a different set of associations and emotions are 'turned on'.

*'I am a passenger on the spaceship Earth.'*

Buckminster Fuller (1969)

A new mindset about Earth was introduced by James Lovelock in 1979 (Lovelock, 1979). The notion of Gaia means seeing Earth as a giant living being.

Lovelock's story starts with the notion, that if you find a sandcastle on a beach otherwise flat or ripple moulded by the wind and the ocean, you will immediately know that life in the form of children has been there.

In a search for a new understanding of life, James Lovelock used the metaphor of Gaia. In Greek mythology, Gaia is the goddess of life – the mother Earth. What he wanted to see and show was the miraculously intricate regulations that make life on Earth possible, to describe and marvel at its ingenuity. For this search, he used the metaphor of sandcastles.

*'When I started to write in 1974 in the unspoilt landscape of Western Ireland, it was like living in a house run by Gaia, someone who tried hard to make all her guests comfortable.'*

James Lovelock (1979)

Sandcastles cannot appear at random or by the normal physical processes of wind and water. Lovelock writes: 'Even in this simple world of sand and sandcastles there are clearly four states: the inert state of featureless neutrality and complete equilibrium (which can never be found in reality on Earth so long as the sun shines and gives energy to keep the air and sea in motion, and thus shift the grains of sand); the structured but still lifeless 'steady state', as it is called, of a beach of rippled sand and wind-piled dunes; the beach when exhibiting a product of life in the sandcastle; and finally the state when life itself is present on the scene in the form of the builder of the castle.'

With this sandcastle concept tool, James Lovelock searched to find phenomena that indicate a situation that is not a 'flat dumb normal' but instead seem to be intentional with the aim of enabling living creatures.

It is, for example, noteworthy that through all of the time of life on earth the temperature of earth has been kept within the boundaries of not freezing and not boiling, both states would have killed all life. Life has developed over thousands of millions of years (approximately 3500 million years). Even if there have been long stretching ice ages, these have been limited events taking place only at the north and south third of the planet. Hence ice ages have not hampered the continued development of life on the rest of Earth. Geological investigations show that through all these years, the climate has always been relatively similar to how it is today and all the time water has been allowed to be in liquid phase as well as in gas and solid phases.

Superficially one may rapidly conclude that Earth is a lucky distance from the sun – neither too far away to be cold like Mars or too warm like Venus and that is the 'luck' of a temperate planet. Earth is located in what astronomers call the Goldieluck zone of not too hot or not too cold – named after Goldieluck finding the good porridge at the three bears. However, in the 3500 million years of life's evolution, the heat output from the sun has increased by 25%. Despite this enormous change, things have remained sufficiently steady for life to continue its evolution.

Another example of the earth controlling its internal states is that the salt content in the ocean has been kept within acceptable boundaries. Salt content above 6% would mean the end of life in oceans as the osmosis process would cause cells to dehydrate. Surprisingly, and through the same period of life development, the salt content has been kept at an average level of 3–4% and never exceeded the zone of dehydration. This in spite of two continuous sources of salt. The first source is salt that is flushed into the oceans from continental run-off, where rivers wash salts out from the ground and transports it to the sea. The second source is salts leaking into the oceans from the hot interior of earth. Together they would cause the salinity of the oceans to reach its current level within merely 60 million years. This is certainly a long time but compared to the 3500 million years of life, some processes must help prevent the ocean from becoming overly saline. Something is removing salt at such a rate that the oceans stay fit for life.

A third example is the air's content of oxygen which is kept constant around 18%; this amount is just below the level where forests would catch fire spontaneously and above the level required for humans and animals to breathe.

These are just three of many examples of regulated 'pumps' that make essential substances available to make life possible. They are supplemented with a multiplicity of other pumps that remove toxic compounds from the biosphere and pumps that ensure nutrients and building blocks for life to occur in a fashion that enables life. These continued 'regulated' movements of materials and changes of material states occur on a global, local and microscopic scale continuously with many different time constants from seconds to millennia. The intricate ways these pumps work is simply amazing. And not only is it fantastic that these pumps of movements exist at all. It is even more impressive that they are synchronised to enable the existence of life through millennia. A lot of the work having been carried out by a multiplicity of life forms, where all are working unconsciously towards that goal of an environment conducive to more life. And these many processes have worked out continuously sufficiently well for life to exist and develop all the while. From a control and automation engineer point of view, that is beyond impressive.

> *'The entire surface of the Earth including life is a self-regulating entity and this is what I mean by Gaia.'*
>
> James Lovelock (1979)

The intricate complexity of the control loops is astonishing, but the wonder does not stop there. Complementary to these numerous loops of homeostasis, life has been able to adapt dynamically as conditions for life on earth have undergone change as a response to life itself. For example, the appearance of oxygen in the world of anaerobic organisms must have been a disaster for anaerobic microorganisms. However, they found a way to continue the life of their species, for example in the guts of living animals and people. Somehow, the Gaia principles ensured a space for all. Lovelock argues that it is – despite the warnings of environmentalists – rather difficult to kill life globally. It is so diverse, and the regulation by this diversity is so robust that earth can absorb quite significant disturbances. Life may regress and in this perspective life on the ground compared to life in the oceans is only a curiosity that might be ended while marine life will continue. In that perspective, human life is even more so expendable for life on Earth. Life itself is surprisingly robust, resilient and sustainable and has been sustained for millions of years. Looking at that time perspective, the sustaining in itself is impressive and worth celebrating. But an even more striking feature is its ability to keep evolving and diversifying.

Looking at earth from this perspective is a lesson not only of sustainability but also of development and evolution. It seems that evolutionary-development and sustainability are in dialogue with each other. That one is the precondition for the other. That changes in the earth system, Gaia, leads to necessary adaptions of life, which in turn leads to changes in the earth system and so on. That is the long term process of life.

Since our human actions have reached a power to affect the global scale, we need to orient ourselves and our efforts to this scale. A realistic principle when trying to perceive the world through the lens of Gaia is that one need to take a very long-time perspective. Native Indians had a leadership principle of working from the perspective of seven generations, i.e. something like 500 years. This is highly relevant and stands in stark contrast to our current western decision-making horizon. Most investment decisions and business cases have a scope of 2–4 years. In utilities, the scope of decisions may go up to 10–20 years, but it is still short and narrow.

It is apparent that an understanding of 'spaceship earth' based on a mechanistic viewpoint is going to miss essential points of how the earth is alive in the sense of a staggering amount of interlocking homeostatic loops – of which the big ones that Lovelock mentions are just a few. Some feedback loops are living and some are not, but seen as a whole, Lovelock posits that earth is one giant significant living being.

The adaptability of the system makes it difficult to predict what will happen, which constitutes one of the major difficulties of understanding global warming to an extent where reliable models can be developed. It seems that the earth has many ways of adapting and absorbing shocks to its system. A large part of global warming is incorporated in the oceans as heat and acidification. Hence it is perhaps more relevant to speak of the increase in $CO_2$ in the atmosphere as an increasing pressure on the earth system, to which the earth system may adapt as long as it can.

> **Your reflections:** If you work with water, you are not a passenger on planet Earth, you are part of the crew. Do you feel competent to take this larger responsibility beyond your job? What do you still need to learn?

## NATURAL LAW AND THE GLOBAL RESPONSE

Oren Lyons introduces a different insight (Lyons, 2004):

> *'I said earlier that my first message to you is that the kind of leadership we have must be changed. The second message I bring you is that global warming is real. It is imminent. It is upon us. It's a lot closer than you think, and I don't believe we're*

*ready for what's coming. We're not instructing our people, we're not instructing our children, we're not preparing for what is coming. And it surely is coming. We've pulled the trigger, and there is nothing we can do now to stop it. The event is underway.*

*What I say to you today is that the ice is melting in the north as we speak, trees are tipping, the roads are buckling, buildings are falling in. From what? From the permafrost melting. Perma. Permanent frost. No, not so permanent. It's melting right now. Four million acres of spruce killed two years ago by beetles. This was caused by global warming, which allowed two cycles of beetles instead of one. The second cycle killed the trees. You can't negotiate with a beetle. You are now dealing with natural law. And if you don't understand natural law, you will soon.'*

Oren Lyons

Oren Lyons comes from the Turtle Clan of the Onondaga Nation where he is the trusted chief, maintaining customs, traditions, values, and history from his clan. Looking through his eyes provides us with a different view of life and the living nature.

Professor of Law, Mary Christina Woods took up Lyons' concept of natural law in her 2010 paper called 'You can't negotiate with a beetle' (Woods, 2010) in which she clarifies how our current 'legal membrane' for the protection of the environment is not working well enough for the society we have now.

*'Throughout most of civilization, human societies have governed their relationship with the environment through a series of codes or rules. Even back in Justinian times, the Roman Empire had legal rules about the taking of fish, ownership of eroded soil, and the cultivation of bees In North America, tribal societies had rules and cultural norms restricting the harvest of species to certain times of the year and prohibiting waste and the soiling of waterways. No matter how simple or complex, all societies create a legal membrane through which individuals act in relation to nature. That membrane is environmental law. The efficacy of environmental law should be of utmost concern to citizens, for any government that fails to protect its natural resources sentences its citizens to misery and perhaps even death.*

*Though most lawyers think of environmental law as just one of several dozen specialties in the law, it is actually a different breed, for one simple reason. Environmental law is accountable to a supreme set of laws – the laws of nature, or natural law, as Oren Lyons and indigenous leaders worldwide call it. The most important function of environmental law is to assure humanity's compliance with nature's laws, all of which ultimately determine whether citizens will survive and prosper, or suffer and perish. If environmental law becomes too detached from nature's laws, or ineffective in assuring humanity's adherence to such laws, society risks collapse – and environmental law, no matter how seemingly complex or sophisticated, will have been irrelevant.'*

Mary Christina Woods

Looking at the global attempt to establish a membrane between human society and global natural law gives a perspective of how slow and difficult change is on this scale.

I was born in 1971. This was just around the turning point for sustainable human existence. Since then, according to the Global Footprint Network, we have been consuming more than the earth can produce sustainably year by year (Figure 4). Like a bank account where more money is taken out than put in, continually causing the balance to go into deeper and deeper negative numbers. At that point, the world at large hardly knew of the global trouble it was getting into.

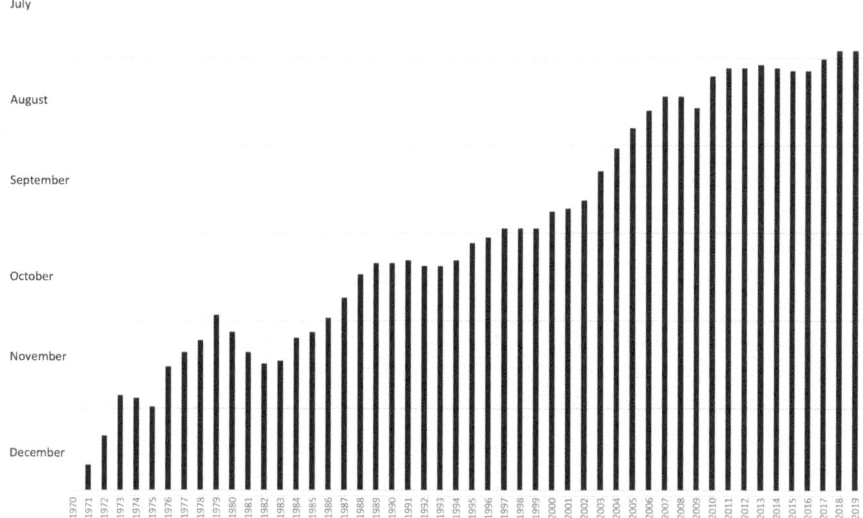


**Figure 4** Earth overshoot day. (*Source*: Global Footprint Network, www.footprint network.org, 2019)

Already in 1972, the first UN environmental conference was held in Stockholm. It was called the 'United Nations Conference on the Human Environment'. The declaration begins as follows:

*'The United Nations Conference on the Human Environment, having met at Stockholm from 5 to 16 June 1972, having considered the need for a common outlook and for common principles to inspire and guide the peoples of the world in the preservation and enhancement of the human environment, Proclaims that:*

*Man is both creature and moulder of his environment, which gives him physical sustenance and affords him the opportunity for intellectual, moral, social and spiritual growth. In the long and tortuous evolution of the human race on this planet a stage has been reached when, through the rapid acceleration of science and technology, man has acquired the power to transform his environment in countless ways and on an unprecedented scale. Both aspects of man's environment, the natural and the man-made, are essential to his well-being and to the enjoyment of basic human rights the right to life itself.*

*[…]*

*Principle 1*

*Man has the fundamental right to freedom, equality and adequate conditions of life, in an environment of a quality that permits a life of dignity and well-being, and he bears a solemn responsibility to protect and improve the environment for present and future generations. In this respect, policies promoting or perpetuating apartheid, racial segregation, discrimination, colonial and other forms of oppression and foreign domination stand condemned and must be eliminated.*

*Principle 2*

*The natural resources of the earth, including the air, water, land, flora and fauna and especially representative samples of natural ecosystems, must be safeguarded for the benefit of present and future generations through careful planning or management, as appropriate.'*

Declaration of the United Nations Conference on<br>the Human Environment (UN, 1972)

In the 1980s the Brundtland commission worked on getting a clearer picture on the 'common outlook and common principles'. The Brundtland Report was published in 1987. It was the first global political vision for re-establishing sustainability and carried the title 'Our common future'. The report warned that we as the human race currently meet our needs in a way that diminishes future generations ability to meet their needs. The task for the report was to: 'to propose long-term environmental strategies for achieving sustainable development by the year 2000 and beyond'. In the 'call for action' it is stated:

*'From space, we see a small and fragile ball dominated not by human activity and edifice but by a pattern of clouds, oceans, greenery, and soils. Humanity's inability to fit its activities into that pattern is changing planetary systems, fundamentally. Many such changes are accompanied by life-threatening hazards.'*

Gro Harlem Brundtland (1987)

In the period 2000–2015, the global community led by the UN worked with the Millennium Goals, focused on eliminating poverty. This worked out to a great extent, not least due to the industrial revolution in China. The eight Millenium Goals (WHO, 2000) were:

(1)   To eradicate extreme poverty and hunger;
(2)   To achieve universal primary education;
(3)   To promote gender equality and empower women;
(4)   To reduce child mortality;
(5)   To improve maternal health;
(6)   To combat HIV/AIDS, malaria, and other diseases;
(7)   To ensure environmental sustainability; and
(8)   To develop a global partnership for development.

At the same time, however, the sustainability crisis kept accelerating, and the sustainable development aimed for in the Brundtland report by 2000 was not reached. Having environmental sustainability as priority number 7 and only vaguely defined reminds us how the focus was much stronger on other types of problems. Key focus was on 'the developing world' and the suffering taking place there.

2015 seemed to mark a very distinct change for the UN as the sustainable development goals (SDGs) were launched. This marks the transformation of the vision laid down in the Brundtland report into goals in 17 areas of the human and ecological system, with 232 quantitatively measurable indicators. The goals are for the year 2030, and the idea is that these SDGs are only the first set of goals that need to be met in order to steer the planet back into the bounds of sustainability. The goal definitions succeed in integrating the three aspects of the economy, the social and the environment:

*'This Agenda is a plan of action for people, planet and prosperity. It also seeks to strengthen universal peace in larger freedom. We recognise that eradicating poverty in all its forms and dimensions, including extreme poverty, is the greatest global challenge and an indispensable requirement for sustainable development. All countries and all stakeholders, acting in collaborative partnership, will implement this plan. We are resolved to free the human race from the tyranny of poverty and want and to heal and secure our planet. We are determined to take the bold and transformative steps which are urgently needed to shift the world onto a sustainable and resilient path. As we embark on this collective journey, we pledge that no one will be left behind. The 17 Sustainable Development Goals and 169 targets which we are announcing today demonstrate the scale and ambition of this new universal Agenda. They seek to build on the Millennium Development Goals and complete what these did not achieve. They seek to realize the human rights of all and to achieve gender equality and the empowerment of all women and girls. They are integrated and indivisible and balance the three dimensions of sustainable development: the economic, social and environmental.*

*The Goals and targets will stimulate action over the next 15 years in areas of critical importance for humanity and the planet:*

**People**
*We are determined to end poverty and hunger, in all their forms and dimensions, and to ensure that all human beings can fulfil their potential in dignity and equality and in a healthy environment.*

**Planet**
*We are determined to protect the planet from degradation, including through sustainable consumption and production, sustainably managing its natural resources and taking urgent action on climate change, so that it can support the needs of the present and future generations.*

**Prosperity**
*We are determined to ensure that all human beings can enjoy prosperous and fulfilling lives and that economic, social and technological progress occurs in harmony with nature.*

**Peace**
*We are determined to foster peaceful, just and inclusive societies which are free from fear and violence. There can be no sustainable development without peace and no peace without sustainable development.*

*Partnership*
*We are determined to mobilize the means required to implement this Agenda through a revitalised Global Partnership for Sustainable Development, based on a spirit of strengthened global solidarity, focussed in particular on the needs of the poorest and most vulnerable and with the participation of all countries, all stakeholders and all people.*

*The interlinkages and integrated nature of the Sustainable Development Goals are of crucial importance in ensuring that the purpose of the new Agenda is realised. If we realize our ambitions across the full extent of the Agenda, the lives of all will be profoundly improved and our world will be transformed for the better.'*

The 2030 Agenda for Sustainable Development (UN, 2015a, b)

Clearly, when reviewing the goals and indicators, they are a compound of many compromises and any organisation that has tried to translate the goals into specific actions has found the process difficult and confusing. This is also the case when we look at water specifically. Looking into SDG 6 on water, the mentioning of integrated water resource management is very vague. Regardless, the SDG is an expression of solidarity across the globe and between generations to the extent that the world has not seen before.

It seems reasonable to believe that in 2030 our understanding has become further refined, and a new set of goals will be set forth for a higher global aspiration. By the invention and perhaps even more painfully difficult ratification of this system of 17 goals with 244 indicators, a new type of global accounting system is emerging. This constitutes an unprecedented level of systemic thinking and acting.

So we are approaching the 50 years anniversary of our human consumption overshooting the capacity of the heart, and we are still balancing on a knife-edge. Will we rise to the challenge or deteriorate into chaos? These are the two competing narratives now. One narrative is a hopeful vision for future in a better world encompassed in the words 'Leave no one behind'. The other narrative is that research shows that if we continue without change, the world will face serious problems, widespread suffering and irretrievable losses.

> **Your reflections:** Most people want to do good. Most people don't want to live in a collapsing world. Most people don't want to leave a damaged world to the next generation. We have known about this for a long time. The world is full of mostly good people. Why can't we get ourselves wrapped around this challenge? Humanity have faced other large challenges, what makes these problems seem so insurmountable?

## THE EMPATHIC CIVILISATION

There is reason for optimism. The mere fact that we can think in terms of globality and that we can collectively effort towards understanding the global system dynamics is impressive and new. Clearly the sustainability insights and thoughts have been going on for quite a while in limited intellectual groups; however, now these thoughts are being thought and taught in elementary school. Even though we are struggling to make enough sense to take action based on them, even if we make flawed conclusions, even if we are bewildered in our judgement of true and false analysis of the global situation, we spend time thinking of it, trying to adapt our brainware to this new mode of thinking. And that is amazing.

What if we only fully can achieve the vision of the SDG goals, if we open our hearts to feel the state of the earth; if we succeed in becoming coherent with GAIA and showing empathy. What if we cannot succeed with our mental capacities alone?

In 'The Empathic Civilization: The Race to Global Consciousness in a World in Crisis' (Rifkin, 2009), an enlightening and charitable analysis of the world development is made by Jeremy Rifkin. He proposes to see a strong human developmental process towards greater and greater empathic reach and depth. This he states is the overarching imperative of human development. It is a difficult journey where history is ripe with examples of daring attempts. The caveat is, however, that while the goal of this humanistic journey is noble, the energy cost is enormous. Hence the empathy quest has burned extensive resources in its wake. Our current situation is historically the most significant attempt to reach the empathic civilization. This is the first time; the attempt is global.

'Empathy' is probably not the first label one would put on the driving force for the process of globalism. However, Jeremy Rifkin provides a long list of examples of how empathy has grown deeper and broader through this process in unrecognised ways. Not because the empathic effects are invisible, more because it is not the focus of our joint narrative, which is rife with conflict and suffering. But next to all the drama a slow quiet empathic evolution has been taking place.

Jeremy Rifkin provides an ample amount of examples on areas in which it is possible to read the development in empathy, be it in the way we raise our children today compared to earlier times or the way we treat minorities or the reach of our acquaintances that often stretch several countries and continents. It can also be read in the overarching themes of historical epochs from great theological civilisations, to ideological ages (18th and 19th century), to the age of psychology (20th century).

It appears that though empathy is not the focus of our attention in the globalisation process 'subconsciously' we are guided in this direction anyway. As soon as we have the ability to reach a higher level of empathy, we go for it. Basically, because it is the way of the 'heart'. It makes us feel good.

However, reaching high states of empathy comes with a cost. The story of the fall of the Roman Empire is illustrative of the problem. The city of Rome was extra-ordinary not only architecturally, but also philosophically. Rifkin writes:

*'From the very beginning, Rome wore two faces. There was the Rome that conquered the world, enslaved millions of people, occupied other lands, delighted in cruelty, and*

*built a stadium – the Colosseum – that could seat 50,000 spectators, who cheered as Christians, criminals, and slaves were fed to the lions. It was also a place where self-awareness grew, individuality began to develop, and tolerance toward other religions became commonplace. … The slave economy allowed urban society in Rome to flourish. But the increased cosmopolitan sensibility was purchased at the expense of an enormous human entropy bill. Millions of slaves were worked to exhaustion and death to provide the energy to run the empire.'*

Jeremy Rifkin (2009)

Similarly, Rome resulted in the degradation of adjacent farmland. Due to the obligation to pay tax, farm-land was over-exploited and lost its top-layer of fertile soil. Wealth poured into Rome from the conquests of neighbouring countries. However, eventually, the wealth of natural resources was exhausted around the empire.

Rifkin writes:

*'The entropy bill was enormous. The available free energy of the Mediterranean, northern Africa, and large parts of continental Europe, reaching as far north as Spain and England, had been sucked into the Roman machine. Deforested land, eroded soil, and impoverished and diseased human populations lay scattered across the empire. Europe would not begin to recover for another five hundred years.'*

Jeremy Rifkin (2009)

We are in a similar process as they were in Rome. But our tools to despoil the environment are more 'effective', but so is our empathic sensibility. So again, we find ourselves at a balancing point between higher self-awareness, empathy and connection on the one hand and a sustainability crisis ready to reverse it all on the other hand.

*'We are on the cusp, I believe, of an epic shift into a "climax" global economy and a fundamental repositioning of human life on the planet. The Age of Reason is being eclipsed by the Age of Empathy.*

*The most important question facing humanity is this: Can we reach global empathy in time to avoid the collapse of civilization and save the Earth?'*

Jeremy Rifkin (2009)

Today's scientific understanding of our sustainability crises is perhaps the change in the ingredient that can tip the experiment towards a better fate than the Roman Empires. The sustainable development goals are an innovative way of handling the crises. It marks a global holistic way of thinking, where if we solve the human social problems and lose the 'nature problem' we end losing, but if we do not solve the human social problem, then we for sure are going to lose the nature problem. It is a departure from a simplistic cause-effect way of thinking towards an understanding that welfare in human society and thriving of nature is connected and we cannot only address one problem and turn our blind eye to the other. The SDG address poverty, hunger, health, education, gender, clean water and sanitation, energy, work and economy, industry and infrastructure, inequalities, communities, consumption and production, climate action, life below water, life on land, peace, justice and institutions and finally partnership thinking.

The SDGs are far from full holistic thinking – especially if one goes into the details with each of the 244 indicator targets, one feels that a lot of important stuff is still missing. Still, it is a huge diplomatic achievement to agree to take such a holistic view on the world crises.

This is a new way of looking at life and looking at earth. Being a water professional in this worldview is a continuation of what came before, but it is also a new beginning. Suddenly, we see the earth as 'Spaceship Earth', which is a departure from seeing nature as civilisation's environment. The word environment comes from French 'environnement', meaning that which is around you. In the new mindset, nature is not 'what is around us'; it is an integral part of our Spaceship Earth. This creates a more tender relationship to nature. What surrounds us take on a different meaning. The water flowing around in a delicate pattern governed by natural law is now seen for what it is, for a holistic whole rather than just as the simpler and fractured understanding of water as drinking water or wastewater, rainwater or ocean. Water can be understood as one big fluid body; shifting shape and content as it streams around the world.

> **Your reflections:** We are balancing on a knife's edge. Can we reach global empathy in time to avoid the collapse of civilization and save the Earth? As polarisation is increasing in so many countries, how shall we bridge this?

## FROM WATER PROFESSIONAL TO WATER STEWARD

In her book, Water: Nature and Culture, Veronica Strang (2004) binds together water in the global, the local and the personal:

*'In this sense, the earth's planetary fluid system is not unlike that of its multifarious organic life forms. It contains some parts with less water content than others, but even in these, water is vital to the successful maintenance of life: all biota depend on the movement of water through air, soil and cells, and all are connected by water. This sense of connection is nicely captured by Vladimir Vernadsky's ideas. In the 1920s, Vernadsky was inspired by early Greek debates about the 'nature' of the earth and its waters, and by Johannes Kepler (1571–1630), a German mathematician and astronomer who – long before James Lovelock resuscitated the Greek notion of Gaia – described the earth as a living being composed of sentient, interactive particles. Vernadsky highlighted the point that not only had all life forms emerged from the oceans to populate terra firma, they remained materially connected by the flow of water between them. This vision of a living, interconnected biosphere was picked up by scholars such as Lynn Margulis and Dianna and Mark McMenamin to describe the 'symbiogenesis' of all flora and fauna, comprising, as*

*the McMenamins put it, a "hypersea" of biota connected by water, in which humankind, rather than flinging out its chest and straddling the earth like a Colossus, is presented more modestly, as one of the myriad species participating in a larger flow of organic life.'*

Veronica Strang (2004)

Following, she elegantly connects this hypersea of all the water flowing in life to an internal hyposea:

*'One of the reasons that it is easy to envision a connective "hypersea" of water linking all living things is that water behaves in similar ways at every scale. In a microcosmic echo of planetary circulations, water flows through even the smallest organisms in what we could call "hyposeas", connecting each part of them. Thus in the human body, as in larger systems, water mediates interactions between all of the different materials and processes involved in maintaining life. And, as in the wider environment, the variability of these materials depends both on their molecular structure and on their water content. Even now, millions of evolutionary years after biota emerged from the oceans, human bodies are approximately 67 per cent water. Human teeth are like rocks, having just over 12 per cent. Bones, which in metaphor serve as the body's timber, are 22 per cent water. Brain tissue, like a fertile, resource-rich wetland, is about 73 per cent, and blood – though certainly thicker than water – is 80 to 92 per cent $H_2O$.'*

Veronica Strang (2004)

In 2017, Pia Soeltoft published a short book called 'Ten things leaders can learn from Kierkegaard' (Soeltoft, 2017), where the central idea is Kierkegaard's powerful idea of 'to will only one thing'. The idea is to clarify to oneself what one's life is about. To decide upon and commit to one clear ideal towards which one keeps oneself responsible, towards which one measures one's daily deeds in big as well as small things. Kierkegaard states that it takes time alone to contemplate and become clear about ones ideal and a number of principles apply for a useful 'one thing'.

First of all, the ideal must be about wanting to do good. It has to stand the test that one wants to live in a world where others might have the same ideal. So, for example, having an ideal of 'living the good life for oneself' would not lead to a good world, because it would lead to a world of 'everyone for themselves'. At the same time, the ideal has to be something that comes from oneself rather than based on other's wishes. Or else it cannot be maintained over time. There has to be a degree of self-motivation. When that is said, this 'ideal' will not be 'a walk in the park', but rather a 'mountain climb'. An ideal is in a sense a continued internal struggle to keep to this one commitment.

However, what is just as important is that while it has to be observable from the outside, it should be based on an inner ideal; not an external goal. Goals have their place, but the ideal informs you on what goals to purchase and how to purchase them, but it never stops. The ideal continues. It is like answering 'what is the overall idea of my life?'.

Additionally, Kierkegaard puts up the following criteria for selecting ones ideal:

- Do I only want my ideal for the sake of the gratification?
- Do I only want my ideal due to my fear of punishment?
- Do I only want my ideal for my own sake?
- Do I only want my ideal to a certain degree?

First, the ideal cannot be for the sake of a kind of gratification, because then it is the gratification one wants. Second, it cannot be to avoid the fear of punishment, i.e. we cannot want sustainability merely from the fear of a grim global crisis. This is not a whole-hearted commitment. Third, one cannot want it for one's own sake only. If it truly is a good ideal it has to be good for others as well. And finally, one cannot only want it to a certain degree because that leads to repeated internal negotiations, postponements and half-hearted attempts, which will eventually water down the commitment to the ideal.

By spending time listening inwardly for this ideal an inner compass is developed … further, I hurry to add. Because it has always been there as we listen to our conscience, as we strengthen and use our will, as we work to know ourselves, as we take ourselves and our work seriously and as we battle with our fears, it is always there: the heart.

By being explicit about this heart way, we become more transparent for others and more transparent to ourselves. This has importance here in the terrain between two stories – the old story that is not satisfying and the new one that we can only glimpse. Let us try to imagine an ideal of what we are striving for. How would that world look? How can we achieve it? We might be so wrong, and we might be surprised by what happens on the way, but we ought to keep this dialogue open and be as explicit as we can be. When and if we realise we took a wrong turn, we have to be open for walking back or taking a new direction.

I believe we need to change the focus of our attention. We have been so occupied with getting enough water for our consumers and customers, but we need to see customers and consumers in another role as well. We need to see them as fellow travellers on this spaceship earth or in this great living being of Gaia and understand that getting water for all their needs and often mindless overuse is not the essential part of our work – though we seem to believe so. The critical part is to make sure that the way humans use water is not causing harm. It is to preserve ecosystems that are still functioning, restoring ecosystems that have been harmed, stop letting pollution reach the natural water and being mindful about the energy sources thereby avoiding damage to the climate of our spaceship. That is our Work.

This is a journey from 'water professionals' to 'water stewards'; a journey we are all on. It means we need to acquire new skills and integrate them with those that we already have. We need to learn integration in so many ways and dimensions. We need to re-integrate our human society with the spaceship again – seamlessly. If we do things from the right heart set and mindset, we can find peaceful solutions to the problems. We can find ways to ensure footprints for all of us as well as the living world around us. But we need to change our mindset to stewardship. We need to care and restore. We need to open up to our feminine side of groundedness in the soil, of care and nurture – beyond sustaining. This is not a matter of addressing only one issue, such as carbon dioxide. It is about changing our ways towards peace, harmony and integration. Everything is connected for better and worse. The story we tell ourselves about the purpose of our job matters. Even if it is difficult to make a change in our system, a change in our story will cause minor changes that will contribute and accumulate to this 'serving of water' as a way of serving life and the living system, that we are part of, that we need, that we love.

This is long-term and large-scale work. It is work where we must learn what it means to be looking seven generations forward. It is work that spans the whole earth, and yet we are each only one person. Therefore at the same time, it is short-term, small-scale work: what can I do today, how can I approach this situation differently and in accordance with the higher aspirations? I do not succeed every time; you will not succeed every time. It will be frustrating. Take care of yourself as you care for the world. Connecting to the heart does not mean saying goodbye to your head, but it does mean a recentering.

Understanding the vastness of the Gaia system and its unfathomable amount of connections and homeostatic feedback loops that keep the whole living system safe and sound is mind-blowing. But it seems evident that there is an order in things, and we all know what is right and what is not. What serves and what destroys.

The Gaia system has an immense robustness, think of the adaptation to solar input as found by James Lovelock. Humans have tremendous robustness and resilience as well, which make them able to live everywhere on the planet. But both have breaking points, and we should not make compete about who breaks first. Neither should we argue about how much additional pressure the systems can take before they are run to their brink of breakdown. That is not a healthy discussion, neither in regards to humans, nor the world system. Instead we need to nurture and strengthen and live our lives in harmony. This means feeling the pain of our current predicament enough to gain the impetus to act and release the apathy. It means taking a new sacred view of life, earth and water. This means we must work on a longer time frame (seven generations), we must understand our close by ecosystems much better. We must see the things that our culture has made us blind to, the extensive evil we do not see in ourselves and the opportunities available in the indigenous understanding of the world.

Lighthouse stories from this research keep me on some kind of track: Emotos' idea of water being a mirror of our inner states of care, gratitude and peace; Berry Wendell's idea that there are only sacred and desecrated places; Rifkin pointing out that the sustainability crisis is about succeeding in the development of empathy; Lovelock's story of the earth as a giant living being; Naess' concept of deep ecology and finally Veronica Strang's notion of a sea, a hypersea and a hyposea.

> **Your reflections:** What kind of sense does a transition of your role from water professional to water steward make to you? What would be your Kierkegaard-ish ideal? How would you contribute where you are now?

I will conclude this introductory chapter with a quote by Veronica Strang, underlining what is obviously true: that we are a part of this fantastic, diverse and beautiful hypersea:

*'Every cultural group has its own music and images, its own ways of reconnecting with water. It is vital that these are cherished, not forgotten in an unthinking, unfeeling scrabble for material advantage. Societies need to remember what water really is, what it means and why it matters. Water is the fluid connection between humankind and every organism on Earth: we are all the "hypersea". The flow of water that animates our own bodies is simultaneously circulating and animating all of the tiny and vast material systems on which we and other species depend. Water is the creative, generative sea that makes and maintains life, and living water is the substance of identity, of the spirit, of the self. We need to replace utilitarian reduction with an appreciation of water as time, memory, movement and flow; as the tides of the heart and the imagination; as the stuff of real "wealth", which is the combination of health and wholeness. With a sense of fluid belonging, through water, it becomes possible to think and act connectively and collaboratively.'*

Veronica Strang (2004)

# *Chapter 2*
# Practical experiments

---

In parallel to the inner sensing, studying, and contemplation on the fabric of water aspirations as described in Chapter 1, ordinary 'life on the outside' continues. In conversation with the reflections of heart and water, the work increasingly becomes a continuous integration of the holistic view into each part – adjusting, listening, designing, testing, understanding of dynamics, mediation, etc.

I work as part of the senior staff in the water utility of Kalundborg in Denmark. This hence becomes the arena for the expression of the inner thought processes and ideas. As well as the other way around, what happens in the utility provides food for thought in the contemplation of water aspirations. It is difficult to convey a one-to-one description of how this dialogue takes place, as in 'I sensed/thought this and then did this or that and then that happened which made me sense/think so and so'. But it is distinctly clear that such a dialogue takes place. And I trust that dialogue to be of vital importance. This is not a work of 'stiff dogmatism', but on the contrary, an attempt to fluidly 'act connectively and collaboratively', as Veronica Strang puts it.

The contemplative experiments of writing this book changed the outcome of both analysis and decisions, but perhaps more importantly changed the perceived options – new possibilities appeared possible. Similarly, experiments with different ways of organising work and people, different types of social processes, different ways of listening etc. changed outcomes.

The dialogue between the inner and the outer circumstances often occurs in leaps. The leaps seem to occur for two reasons or in two different kinds of processes.

© IWA Publishing 2020. Water Stewardship
Author: Pernille Ingildsen
doi: 10.2166/9781789060331_0053

One process is when an insight initiates an inspiration for change. In that case, it takes time to translate the idea into real actionable tasks. Second, it takes time because a new way of doing things in a utility most often needs to be anchored with a number of people. Sometimes it takes time to convince people to take the kind of risk a new idea constitutes. And in that discussion, many people will come up with reasonable objections that need to be addressed. There is no point in forcing it through – it will increase the risk of failing dramatically. So it takes time and often lots of feedback loops and adjustments. When it finally works out it seems as if it happened in a leap.

A different experience is when a new pattern emerges out of what happens in daily work in or around the organisation of the utility. What happens may either dissonate or resonate with the contemplations. It is as if due to these contemplations new things become visible and cause surprise. The myriad of things that take place in a utility organisation become a springboard for understanding the new mindset in a new way, generating new insights and ideas. Again, seeing a pattern once or twice is usually not convincing, there has to be a cognitive recognition – which often takes the form of a 'eureka moment'.

I have chosen to write about my work at Kalundborg Utility to illustrate how we have found practical ways to move in the direction of increasingly taking on a water stewardship role. I will try to explain how I have interpreted it, how we have experimented and what types of changes this has brought about.

I began in the utility at the same time that a new CEO had been appointed. This marked a significant change in management of the utility and a change in strategy towards a clear strategic marker of 'sustainability'. The change process in Kalundborg Utility is therefore also a case story that gives an example of a trajectory of a journey towards sustainability in a water utility. This is NOT an attempt to give directions or even inputs to how this journey should be made. Instead, it is an attempt to root the ideas and the whole endeavour of water stewardship in practicality and common real-world projects and issues. If it is to be taken as an encouragement, it is an encouragement to experiment, to look at these experiments and then design better and more daring experiments.

What I have tried to make clear in the below cases is what we managed to 'upgrade' from project to project. So it is not about doing the ultimately right thing – though every time we tried. Actually, what is a richer focus is to look for what was upgraded. Later as this focus became clear, the question changed to: 'what do we want to upgrade?'. This is particularly clear in the last case below.

## COLLABORATIVE DEVELOPMENT OF A NEW UTILITY STRATEGY

Before we started working with our five-year strategy towards sustainability, the top management worked out a set of values from which leadership in the utility would spring in the future. The chosen values were: passion, relationships, ambition and motivation and they set the tone for the strategy to come. These values were an attempt to change from a culture of disengagement, static background dissatisfaction, stress and a lack of direction and imaginative courage. With the new values, a new state of mind in operating the utility was called upon.

The values spurred a great deal of controversy both in our utility and even in some of our 'fellow utilities' because the values marked a stark departure from the traditional utility values which can be characterised as operational robustness, conservativism and risk-avoidance. The new values stated a different way to be a utility. Some greeted the change as a breath of fresh air, others became anxious about the meaning of this for themselves and the work they mastered, and still, others felt angry – either because they thought the earlier values were now deemed wrong or because the new values were too far from their own values and competences.

The process and choice of the new values were carried out solely by the top management group of the utility. In retrospect, I think that this was needed to change the agenda so markedly, to communicate the radical change clearly. But it was at the same time evident that this was the kind of top-down approach that we actually had to change away from. If we were really to achieve a future where these values would be authentic, this could not be the way to move forward in the future. Therefore the following strategy process was designed to be distinctly participatory. The strategy process lasted a year and was carried out in the following steps:

(1) Brainstorming;
(2) Grouping of themes based on emerging patterns;
(3) Internal and external analysis;
(4) Deciding on strategic markers;
(5) Translation of strategy into actions.

The brain-storming phase took input from each line of utility (drinking water, wastewater and district heating). Together we tried to answer what needed to be done or with a different twist: what would be good if it was done. For the area of, for example, wastewater treatment, there were close to 100 ideas for activities and concepts to be implemented. The ideas were of different quality and some almost contradicted each other. The ideas were applicable on different abstraction levels, many being neither strategic nor directly actionable. So it was really a messy body of ideas. In the group, we tried different kind of groupings which revealed some redundancy in the ideas. But even redundant ideas were formulated differently by different people and hence giving a different flavour to the idea and just merging the ideas into one headline seemed to reduce or dispirit the idea.

Eventually, an organisation of the ideas with a strong explanatory power was 'invented'. The ideas were grouped according to the mindset the idea represented. First, we saw that the ideas were an expression of three different focuses: (1) cost reduction, (2) quality improvement and (3) customer service and participation. Later this was extended with two additional mindsets: (4) environmental concern (later renamed sustainability) and (5) innovation see Figure 5.

The five categories became an important inspirational and organising principle. The method inverted the problem to: 'If our strategic direction was mainly 1, 2, 3, 4 or 5, what would the main actions in our strategy look like?' This caused five different futures to be imagined and described. To support our imagination and memory, we linked a company name with each scenario, a company name which seemed to be governed by either of these foci. The cost focus (1) was represented by a Danish low cost supermarket retailer (Netto), the quality focus (2) was represented by a luxury car brand (Porsche), the customer focus (3) was represented by a customer-centric mobile phone designer (Apple), the environmental focus (4) was represented by a Nordic environmental product label (Nordic Swan) and finally the innovation focus (5) was represented by an American innovation consulting company (IDEO). It was not that we wanted to copy these companies, but it was a very nice 'short-hand' way of speaking of the directions in a way that everybody understood each of the five general directions.

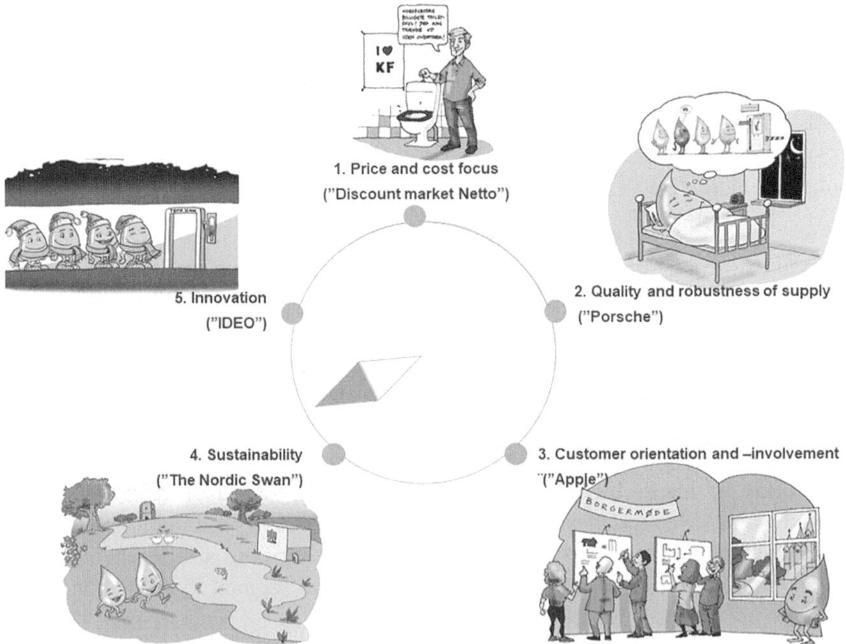


**Figure 5** The five strategic directions identified in the strategy process. (*Source*: Kalundborg Utility)

After describing the five different futures (including vital technical aspects) an analysis was undertaken with five key questions: What do our private customers prefer? What do our industrial customers prefer? Would our key stakeholders support this? Is it economically feasible? Is there anything legally preventing the strategy?

The analysis of the first questions was carried out by focus group interviews. In the focus group interviews, different types of private customers were presented to the five strategic options. The analysis revealed a few surprising points. Across all three customer focus group meetings, a similar pattern played out. When first presented with the cost-focused strategy option, the group loudly applauded, this was what they wanted: lowest possible cost! Following, when presented with the high-quality focus, they changed their mind. After the presentation, this was what they would rather have! Some exhibited the same pattern when presented to the customer involvement direction, though here there was a difference between customers. And finally, there was very high support for the sustainability focus (in this context the innovation-focus strategy appeared more as a tool than a strategy relevant to customers).

The overall conclusion was that especially customers in the main city centres wanted quality at a reasonable cost and would be happy to pay extra for initiatives with a visible positive effect on the environment, preferably in the form of beautiful waterways to be seen in the city. For the industry, the cost is always important, but within reason, and it was evident that sustainability and robustness of supply were higher on their agendas and sparked more energy.

Finally, when the analysis of the five key questions had been carried out, all management personnel were gathered for a 2-day workshop to present and be presented to the results of the analysis on each of the three main utility branches: water, wastewater and district heating. It was clear at the workshop that deciding which strategy to pursue was an abstract and challenging exercise for many; some had strong opinions while others seemed more confused and undecided. This provided a new problem. We wanted a high level of organisational involvement, but how could we broker these different opinions in a responsible, fair and transparent way?

At some point, the top management group took a leadership decision. The group had a separate meeting to have a debate based on the input they had received, and then they would take a vote as there were also conflicting viewpoints within the top management group. Based on the vote it was decided that sustainability was the primary strategic marker and innovation the secondary one. The ambition was to use innovation as a tool to make sustainability come about without excessive additional cost.

When the organisation was presented to this decision, some were very outspoken against it. There was a confusion of whether that meant that we as a utility didn't care about cost, quality or customers anymore. As a top management group, the standard procedure of operation would be to quiet down the critique in a more or less hard-handed way – to show 'leadership strength'. But this would be the exact opposite of the way we wanted to move the organisation.

Instead, we found a way to improve the strategic storytelling to include the 'opposing viewpoints' as well. In this discussion we realised that throwing overboard the traditional utility values was neither what we wanted nor necessary. Instead, we realized the different directions were rather different layers of the story of the utility see Figure 6.

At the beginning of the utility's lifetime it had been about establishing a system with operational robustness. Later, when the system had been running for decades, 'cost and pricing concerns' came into focus. It was probably less than a decade ago that customer service had become an additional theme in life in utilities. Hence the strategic story became a recognition of these foundational values at the same time as it added a new layer of sustainability and innovation. And it was apparent that there had to be elements of development for all five directions. From a strategic point of view sustainability and innovation were the new themes that had to be incorporated in the utility culture. It was clear that on occasion this layer would be in conflict with the more inner layers and at other times they would enhance or upgrade them.

In a sense, one could also argue that the sustainability dimension had been there all the time in the form of compliance to environmental law. However, in the forthcoming strategy period, we were going to 'strengthen that muscle', to understand our effect on the environment better and to find solutions that take environmental issues better into account, than what we had done in the past.

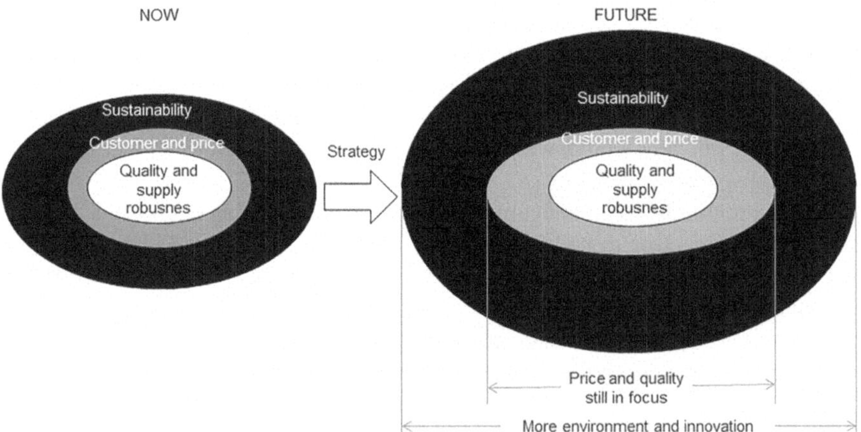


**Figure 6** The new strategy is not a radical departure from old virtues of robustness, quality, economy and customer focus. Instead, sustainability is a new layer we are adding to the work. (*Source*: Kalundborg Utility)

So in that sense, all strategic directions or themes were integrated but with different approaches and with different emphasis. What the strategy succeeded in doing was changing the discussion and, naturally, including more elements of sustainability and innovation in our work moving forward.

As every strategy book in the world states, one thing is defining a strategy, implementing it is a whole other and different matter. We did a lot of interesting work in terms of strategy implementation; some worked – some didn't. But what is important to note is that the strategic story became the foundation of a lot of the following work and is a basis of the case stories that follows in this chapter.

## What was upgraded?

### *Training in providing processes for complex tasks*

Retelling the story makes the process look nice and more well-planned than it was. At the time, it did not feel so. However, it worked well that a general process was planned and visible to all. It meant there was a direction and steps to complete. We had help from a strategy consulting company to support the process. A surprisingly difficult thing was to make sure that everybody knew what the word 'strategy' meant. This was difficult to many, who did not see that there had ever been a strategy to the work they were doing.

### *The interview-translate method*

It was also difficult for many employees to carry out some of the work packages. The work seemed too abstract to several of the key personnel in the utility. To overcome this, we changed the work method, so that instead of formulating strategic analysis themselves, they were interviewed and their input was transformed into statements useful for the strategy development. Every time we read the statements back to the specialists for confirmation that we understood it correctly. This method became a standard method in many subsequent discussions and projects. The issue that a lot of knowledgeable persons do not have a strategic mindset had to be solved to get all relevant knowledge up for decision. The translation of tacit or silent knowledge into defining parameters for decisions became an important method in many decisions to come.

### *Sustainability on the agenda*

The strategy process put sustainability on the agenda in the utility. The word sustainability has been around since the Brundtland report and is often used in the media. However, it is clear that there is still a long way from having heard the word to really understanding it and even from there to translate it into actions. In the utility this was the first step in a long and winding journey towards higher degrees of sustainability.

> **Your reflections:** Do you recognise these layers in the story of how we work with water? To what extent can it be both-and and where does it have to be either-or? What advantages and disadvantages do the both-and and either-or thinking and acting have?

## EFFECTIVE SUSTAINABILITY THROUGH A PROJECT MODEL

Implementation of a project model was of paramount importance to the strategy. The project model facilitated both a structured movement towards sustainability and an improved framework for innovation. At the same time, the project model strengthened a process of increased project management professionalization, an increased focus and an increased clarity on issues like cost, operational robustness and customer involvement. The project model meant a strengthening of the overall project culture. The starting point was that projects were carried out intuitively supported by a few contract templates and checklists. There was a difficult gap in the organisation's ability to align the projects with any strategy. The problem was apparent on two levels: the strategic portfolio prioritisation level and the project execution level. Many projects only exchanged one piece of infrastructure with a new piece following the same design. There was a lack of a critical and constructive discussion of how to upgrade or change the system. And there was hardly any discussion about how to judge whether there was a reasonable relationship between the cost and benefit of each project.

Implementation of the project model was carried out in order to align the project processes and ensure a methodological approach to key questions.

The project model was designed based on the principles of stages and gates. That means that every project timeline is divided into the number of phases divided by gates. For each gate, there would be a gate meeting, where the approval of a project owner or steering group was required to proceed from one phase to the next. Each project hence has a project manager and a project owner/steering group. In the gate meeting the project manager reports the results of one phase to the project owner. The project manager has to present the stage of the full project for 'the gate to be opened' and the project to move on to the next phase. In Kalundborg Utility we used the project model shown in Figure 7.

The purposes of each phase are:

### 0 Proposal

This 'phase' is meant to collect all suggestions and inputs from the whole organisation regarding what projects need to be carried out in the utility. At portfolio meetings the proposal list is prioritised, some projects may be

| 0. Proposal | 1.  Framing | 2. Concept | 3. Design | 4. Implementation | 5. Evaluation |
|---|---|---|---|---|---|
| Description<br>Purpose<br>Initial busines case<br>Project estimate | Ambitions<br>Goals<br>Success parameters<br>Scope<br>First budget<br>Key risks | Concepts<br>Innovative elements<br>Comparison<br>Business cases<br>Evaluation<br>Sustainability<br>Recommendation | Detailed design<br>Tendering<br>Permits<br>Evaluation of<br>succesparameters<br>Budget<br>Risk assessment | Tender process<br>Negotiation<br>Choice of supplier<br>Building<br>Budget follow up<br>Tests<br>Operationalisation | Project evaluation |

**Figure 7** The project model (in brief). (*Source*: Kalundborg Utility)

rejected, the rest are prioritised. In the beginning our focus was on whether a project was to be carried out or not. However in time we found out that the question of *when* rather than *if* it was to be carried was more relevant. The question of *when* was determined using the mindset of asset management, risk analysis and business cases to make an evaluation of whether the lifetime of the asset can be continued or if it constitutes a critical risk to the robustness of supply. This often requires an assessment of the probability of a threat, consequence of a risk as well as determining the relevant backup plan. Carrying out this kind of systematic risk assessment provides an analytical and objective way of prioritising projects.

### 1. *Framing*

Framing of the project is crucial for the success of the project and for ensuring mutual understanding between project manager and project owner/steering group about the aims, limits and means available. Projects can be framed in many different ways; hence if this is not done openly and consciously, everybody will make up their mind sub-consciously – and in different ways. This may cause a lot of confusion and frustration during the process, and it often leads the framing to become a matter of continuous debate or conflict. Important results of the framing are scoping of the project, defining deliverables from the project, defining success criteria and level of ambition and it includes preliminary time plans and budgets.

### 2. *Concept*

The concept phase is crucial for innovation. In an organisation where innovation has played a very small role, we started off with small changes. In the beginning each project was required to suggest at least two different concepts for solving the project challenge. It was surprising to realise that for many project managers it was difficult to find two different solutions and for some even to see why that would be relevant. Moving one step back and truly understanding the project challenge, understand its purpose, rather than just seeing a solution, is a surprisingly tricky mental movement for many people. This caused a pedagogical dilemma to be solved by the project owner. They had to insist on more than one concept – sometimes against a fair amount of resistance or confronted with poor alternatives. At the same time it was a competence development process where the project owner had to find a balance between being supporting and facilitating and being insisting and appropriately critical of poor concept development.

### 3. *Design*

The design phase is generally where engineers and technicians feel most at home. In this phase, we often found that support for risk assessment was important. Again, the challenge had to do with culture. It is not an intuitive act to separate probability from consequence and to address these two different aspects separately. Additionally, a lot of water professionals have a very conservative view of risk, such as 'if there is a risk, it must be removed'. So this is also something to train. Another way to challenge the design is to see if additional functionality can be included in the design. Designers are often focused on meeting minimum requirements only. But fulfilling additional 'wishes' is where we can draw new pride and happiness in the solutions – and done at the right time it does not need to be expensive. It is a challenge of trying to be able to integrate more functionality into the solutions. And thereby meeting more needs.

### 4. *Implementation*

The building phase is the busiest phase. This is where proper preparation meets reality – or not so seldom, where lack of preparation meets reality. Small mistakes or small issues that have been overlooked can have huge effects at this stage. When a project is 'bleeding' it is often the innovation that is cut from the budget first. Therefore it is important with diligent project management to respect all the mundane tasks of project management and inclusion of experts from all relevant technical fields – in the preparatory phases as well as in the implementation phase. The better and deeper the project manager knows the project design, timeline, risks etc. the more able one is to react 'on one's feet', when mistakes occur. Still, one should be very careful to respond too fast. Reflection is the antidote to mistakes making snowball effects.

Another important aspect to plan carefully for is the transfer of the project to operations – this is a topic that could warrant half a book on its own. If the operation department has not been tightly knit into the project team, this is where a whole new set of problems begin. Additionally, to really succeed in establishing a future respectful and sustainable operation of the new system, where proper care is paid, it is important to have everyone involved – including operations. They need to play an active role during the project. There is often a tendency for conflicts between construction engineering and operation. It requires that both groups bend towards each other to work things out. This again requires a safe space for this work as well as genuine respect and mutuality.

### 5. *Evaluation*

So often skipped, with loss of reflection and experience as a consequence, the evaluation phase is important. The times when we have put real engaged energy into evaluation are the times when we have learned the most. And there is really no way around this. Of course you learn as you progress through the project, however in 'the heat of the moment' you do not see as clear as you believe you do. Reasons for frustrating mistakes, behaviours or events often do not appear or dawn upon you until during reflection and seeing the situation from more viewpoints later on. Interestingly, most evaluations end with similar conclusions: 'we need more dialogue', 'we need to address the hard problems head-on and with all involved around the table as soon as possible', 'we need to build more social capital early, problems will wear on that capital and if there is not enough, we will all suffer from anger towards each other', 'we need to communicate eye to eye when conflicts arise – not by email, though it seems more comfortable in the moment' etc.

## What was upgraded?

### Improvement of project management skills

Project management is a separate field of competence, and most technical staff are not sufficiently prepared for this kind of work. In spite of the fact that most water professional engineers will carry out lots and lots of projects over their lifetime. They only get better if project management is elevated to its own field of expertise. It is as if the educational system imagined that all technicians and engineers were going to work under a skilled project manager. This is not the case, most will have to lead projects themselves. And the social skills and agility required to 'survive', or better yet to excel, in project management need to be trained systematically on the job. It becomes better with experience only if there is a focus on it. If there isn't a theoretical framework and an operating process framework, improvements in project management skills will be slow or lacking. This will have serious consequences for the ability to implement sustainability, because taking in regards to sustainability is a matter of integration of increasing numbers of aspects, which means implementing solutions that are often more complex – or perhaps, even more precisely, requires a higher complexity of mindset to find the simple solutions that satisfy many needs and requirements. Therefore, being able to carry out a project with a complex solution is not straight forward. It requires project management skills.

### Project ownership is not trivial

Having the phases working and the project management skills in place are the first stepping stones for an effective, competent and professional project department.

However, to excel in the role of the project owner is important. In many organisations project ownership is done with inadequate focus and competence. There seems to be the idea that this is an easy part of the management job. That perception gets in the way for improving performance in project organisations. The ability of the project owner to challenge the project, while ensuring the sense of a safe space, is quite difficult if one hopes to improve the project. The theory of 'Secure Base Leadership' is useful as a theoretical framework for the work as a project owner.

### *Application of Secure Base Leadership*

The fundamental idea of Secure Base Leadership *(see, for example, Kohlrieser (2012))* is that the leader needs to develop a relationship with the individual project managers in a way that creates a secure bond between them. This is the responsibility of the leader. To have a secure bond means that the leader needs to act in an accountable way; that the project manager can rely on the person and that the leader take a special interest in developing the bond between the two. From psychology, it is well known that secure bonding is of primary importance for a life led with ease, happiness and a sense of meaning and inner balance (see attachment theory for more details). Originally this was found to be true between mother and infant, but recent research has found that a similar principle is relevant in the workplace. When a leader succeeds in creating a secure base, the project manager will feel safe, accepted as they are and through this will be inspired to live up to expectations as they evolve. This requires empathy, common goals and daring to trust one another. Relationships in this context are continuously built-up, maintained, degraded and rebuilt. This work has to come from a 'caring and daring' perspective and we need to identify and realize when we act from fear to find ways to avoid this. This also means that the project owner or steering group should encourage the project manager to take risks. In many utilities risk-aversion is so ingrained in the mindset that it takes a conscious and uphill struggle to loosen its grip and bring new life. This is not to encourage a risky behaviour per se, but rather to understand when minor risk stands in the way of more progressive solutions. This of course requires the leader to be mindful rather than blaming when risk occasionally materialises.

### *Portfolio management is the key to making an effective impact*

Many utilities become nothing but asset management units, which maintain and refurbish the system as it is. This in itself is challenging to carry out in a cost-effective way. However, this is where the importance of strategy comes in, to make sure that there is a direction in the work – a higher aspiration, that at least to some extent transcends the pure cost-benefit calculation.

## RESPECT FOR WATER, TISSO II

In 2019, Kalundborg Utility won the WEX Global Circular Economy Water Project of the Year. The prize was given for developing a full-scale water treatment plant (Tisso II) to deliver water of drinking water quality based on water from a lake without the use of chlorine.

It took slightly more than five years to finalise Project Tisso II. The project is a good case to investigate what worked and what did not work in terms of translating the strategic direction of sustainability and innovation into a large scale project. The project clearly shows that sustainability and innovation do not consist of singular events, but an ongoing process requiring collaboration, focus and a good portion of grit.

Project Tisso is an example of a project that started on the wrong foot. Due to lack of experience with defining a new water plant, we had given too much defining power to external consultants, consultants with which we increasingly came to disagree. It developed into a series of conflicts over approach, values and choice of technologies. Our relationship became increasingly sour as the project progressed. And as the conflict escalated, we slowly but steadily moved upwards on the stairs of conflict. I had times of great inner struggle; because as a project manager I deeply wanted the project to be one of generous and effective collaboration. But we felt there was a lack of reciprocation as well as a lack of listening and attention. Internally, we had many contemplations of what the reasons for this might be and how to solve the problems. The 'collaborative approach' ought to be working, we thought, but we could not make it work. And this wrong start kept haunting the project again and again as we moved forward and all the way to the end of the project.

My interpretation today is that the consultant's and the utility's project team were too far away from each other mindset-wise, and because we were not sufficiently precise in what we wanted and perhaps not clear enough in our communication. Hence large and small conflicts became the order of the day in the project execution.

At some point, an additional consultant company was included in the team via a public–private innovation partnership to take care of the process and machinery part of the water treating facility, while the original partner's responsibility became limited to the building and architecture.

A facilitation team of two people were included in the project to attempt to 're-establish peace' in the project. The facilitation team included an internal project member who was being taught facilitation and an experienced facilitator. To get on to a new start the newly enlarged team met, and a new collaboration model was established. The team now consisted of the internal team, the two external consulting groups and the facilitators.

It was a feeling of great relief to have a joint whole day session on co-developing the new 'Manifest of Good Collaboration' and having this sense of re-alignment in aspirations and direction. The new collaboration model kept the group together and the project moving forward. However, despite the efforts, the project cooperation relationship remained difficult throughout the design part of the project with a number of tough low points.

Had there been only one reason for this, this could probably have been rectified but the reasons were multiple, complex and opaque as several of them had to do with difficulties in the partner's own organisation. It did not have to do with lack of competence – though at times it came across like that. It did not have to do with the people being 'bad or unlikeable', though at times we felt that way in frustration.

What it did have to do with was: (1) a lack of ability, resources or will to deeply engage with the challenge and the aspirations; (2) too much distraction from other projects leading to a lack of attention to both the detail and the whole totality of the project. Both types of attention are necessary, and it is very demanding to hold both; (3) A lack of attention to the psycho-social field in the project. That is having a sense of when to step forward, when to step back, when to challenge, when to work to increase trust, when to explain, when to be prepared with a plan of actions etc. Again, this comes back to the lack of a 'sensing attention'.

The problematic issues were addressed and collective solutions were found again and again, but at some point it had become such a sore point and patience was running short. Therefore, the partner's role was reduced to an absolute minimum. The relationship to the partner changed from a focus on collaboration to one of culpa and repairments. A negotiation battle continued at director level for years and did not reach the end until after the plant had delivered water for months.

In the learning phase a workshop was arranged where representatives from the partner attended. Still the underlying reasons for the partner's predicament never became clear. But clearly it had been a negative experience for both the project and the consulting company – neither benefitted; both suffered in all the ways one can suffer in a large project including economically, reputationally and stress-response wise.

Why speak about this failure in the project when the project as a whole – in spite of these obstacles – today stands out as a world-class innovative and sustainable water treatment plant? For several reasons. First of all, to convey the important message that working in a new sustainable way is not 'a walk in the park' – it is fundamentally difficult, and having a sustainability focus or a focus on 'good collaborative processes' and frameworks is not always enough. One reason for this is that the new mindset of collaboration and sustainability may run smack into the existing mindset and compatibility is not always found. I would not say that we learned enough to avoid similar situations in the future. However, there are a lot of pitfalls we will not fall into again. The keywords are still dialogue, trust-building, rapid conflict resolution, secure base leadership, attention, understanding both the whole and the details and taking responsibility for defining the project frame including purpose, success criteria, ambition levels and correctly identifying needs, requirements and wishes etc. However, it is also paramount to make sure that the partners you engage with have a mindset that is strongly entrenched in collaboration. This can be tricky in the current competitive environment and it requires contracts that can handle this.

So let's change attention into something that worked out well. The advantage of not being too experienced in building water treatment plants was that the utility could muster the courage to set high aspirations in regard to innovation and sustainability per the utility strategy. An advisory board process was set up to help meet these expectations. This process resulted in a solid innovation catalogue and clarity in stating the innovative story and purpose of the project. The advisory board process worked as a kind of spinal cord in the design phase of the project. It unified everybody, it gave a clear and concrete goal to work towards, and it inspired everybody to become inventive and willing to take innovative risks.

The participants in the advisory board process included all internal stakeholders, a representative from the water customer, the project consultants described above with some of their c-level managers and a group of three Scandinavian professors. A skilled facilitator helped design the workshop and facilitate the process.

## Workshop 1

At the beginning of the advisory board, we defined the purpose of the process as this: 'The purpose of establishing an advisory board for the Tisso II project is to ensure a high level of innovation and cooperation; innovation that will ensure lower cost (capital and operational), safer operation and smaller environmental footprint of the plant. At the same time, the plant will work in the future as an integrated part of the campus that Kalundborg Utility is building up, a campus platform where students, scientists, practitioners and industry can test new ideas, equipment and processes in a full-scale environment in a way that ensures safe operation.'

The process was started in an open-ended way without a well-defined end-result in mind. The open-ended approach gave the process a feeling of 'coming together', where everybody had an opportunity to contribute with what felt important. And that was precisely the idea. Not to limit the outcome by the limitations of the project manager's ideas, but to let all internal and external stakeholders and experts voice their thoughts and ideas.

So while we grappled in the first workshop with trying to get everybody to understand the many aspects of the challenge and at the same time guide the innovative impulse, the participants in return inspired us with something we could not have come up with on our own. We went through five themes for the day: the project organisation, the plant design, control and automation, membrane technology and the future campus. And at each theme we were asked difficult and sometimes uncomfortable questions by the panel of professors and customers.

At some point while trying to explain the possibility of protecting the water body from pesticides as opposed to treating water to remove the pesticides afterwards, our Norwegian representative made a joke about the Danish water mentality and said: 'you Danes have an almost religious relationship to your water'. While this caused everybody to laugh it also caused some debate and reflection because there is some truth to it, but the reflection was also that what we really wanted to perpetuate in this project was the concept of 'Respect for water'. Hence 'Respect for water' became the overarching headline for the whole project. A headline that we kept returning to as in for example: 'Does this give respect to water?'

Another surprise was the Danish professor who, exhausted after a long day, said: 'I am surprised that up until now nobody has said the word 'sustainability'. It is like it is in the air, but we are not treating the topic of sustainability deliberately'. And yes, this was really surprising. It was 'in the air' and yet it was true that all the other issues around the plant had a strong tendency to take priority. These two inputs became cornerstones in the work moving forward.

## Workshop 2

In the second workshop 'respect' was at the centre stage in all presentations. The workshop was structured around four key stakeholder presentations:

- Respect for the experiences from the existing waterworks – by a plant operator
- Respect for the users of the water – by the main water consumer
- Respect for technology – by the technology provider
- Respect for the lake and its streams and rivers – by an external consultant carrying out a measuring campaign in the lake

Each presentation brought something of interest to the project. The group was divided into three groups. In each group, one person was assigned the role of being a 'pink duck'. The responsibility of a pink duck is to bring in an 'appropriate disturbance' into the group by whatever means he/she can come up with. Especially, if the group tend to agree too quickly, the pink duck should play the role of the 'devil's advocate'. The project ended up with seven key themes for innovation in the project see Figure 8.

The themes were:

(1)   Symbiosis and sustainability everywhere
(2)   Perfect traceability of water
(3)   Disinfection without chlorine
(4)   Predictive risk management and control

(5)  Super integration of hygiene in the design
(6)  Good manufacturing practice
(7)  Stringent and logical design.

The presentation called 'Respect for the experiences from the existing waterworks' made by the operator had some important implications both on the day of the workshop, and also in the continuation of the project. The operator was not accustomed to doing presentations – especially presentations to professors – and was hence understandably a bit nervous, but nevertheless he was ready to meet the challenge. We decided that I, as the project manager, would help him with the structure of the presentation, but that he would be the one to provide the content. Having this dialogue about what he wanted to say and my different suggestions on how this or that could be presented was for me a valuable insight into his world. It was an 'exercise' that forced me to try to grasp that knowledge more deeply.

We came up with a nice presentation, where he presented three things he liked about the current plant and five things he disliked and would like to see resolved in the new plant. In headlines this was:

**Likes**

(1)  Being included in the project from the beginning to really get to know the processes.
(2)  That the processes are visible with the naked eye.
(3)  An additional screen to extend the view of the SCADA system.

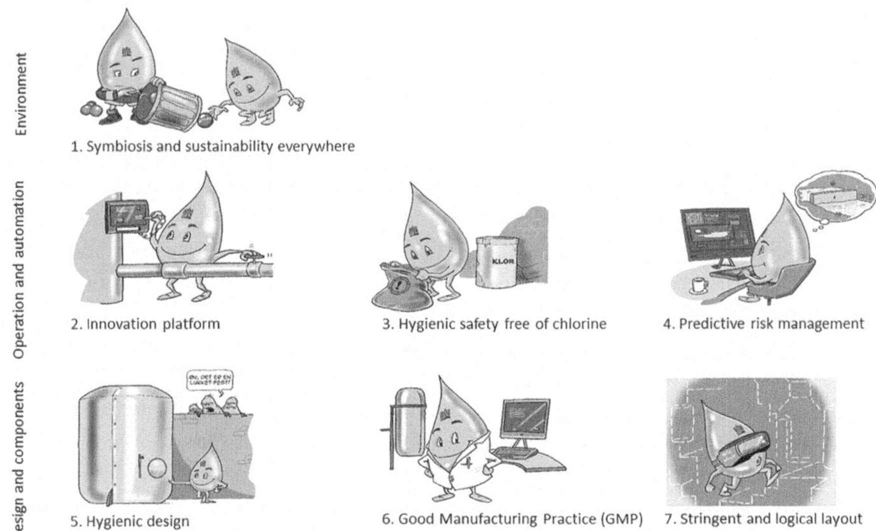


**Figure 8** Innovation themes in the Tisso II project. (*Source*: Kalundborg Utility)

**Dislikes**

(1)  There is too little space in the building. Things happen on different levels on top of each other.
(2)  The operation is made difficult with the lack of panels and lack of ways to change the process by, for example, changing settings, controlling valves or pumps etc.
(3)  Chemical storage is too small and we need to order in too small quantities all the time, which is expensive.
(4)  Choice of material – some of the building material cannot withstand the chemicals.
(5)  Supply during break-down is always critical.

Everybody was very interested in understanding the reason and the consequences of these likes and dislikes and everybody was happy to hear a voice that is often not so loud at this stage of the project. For the operational department this presentation also marked a change to greater engagement and activity in the project. It also helped bridge some of the operator–designer conflict gaps, by having something specific to talk about and work with.


## Workshop 3

The third workshop was carried out like a game, that had both collaborative and competitive elements. The participants were divided into two groups with a mixture of the different roles in the project. The game had seven sets – one for each innovation theme identified in workshop 2. Each set was played the same way. An expert presented ideas on the topic for 10 minutes, then the workgroup shared their ideas on the topic for 10 minutes. Then each group had a dialogue to find good ideas related to the topic based on their diverse backgrounds. Then each group was given 10 minutes to present their ideas. The group was given one point per idea. Of course, this could have potentially led to idea-myopia, so it was stressed that the ideas should be shaped in a way that they could give the project what we called 'an interesting success'. Additionally, it was stressed that the project already had enough work on their table, so ideas should rather have a feel of 'work smarter not harder', than the opposite.

The work resulted in almost 100 ideas. These were scored by everybody individually based on the criteria: resource intensity (low resource consumption was a good thing), price (preferably low price) and height of innovation (innovation height was good). Based on these parameters the final innovation catalogue was prepared with innovation activities within each of the seven innovation headlines. A modest prize of a couple of bottles of red wine was given to the winning team; together with the honour – which probably was more important.

| Innovation | | Innovation | | |
|---|---|---|---|---|
| 1. Symbiosis and sustainability everywhere | 1. Degassing replaces chemical (NaOH) ✓ | 5. Limited access to the waterway ✓ | 5. Hygiejnic design |
| | 1. Energy recovery (turbine) ✗ | 5. Easy to inspect ✓ | |
| | 1. Operational optimization towards sustainablity KPI's ! | 6. Perfect traceability ✓ | |
| 2. Innovation platform | 2. Reference test line ✓ | 6. User requirement specifications for components ! | 6. Good Manufacturing Practice (GMP) |
| | 2. Smart Machines ✓ | 6. Green card for chemicals ✗ | |
| | 2. Three-phased startup ✓ | 6. Open for customer audits ✓ | |
| 3. Hygiejnic safety free of chlorine | 3. Early ozonation ✓ | 7. Functional decomposition ✓ | |
| | 3. Biostability studies ! | 7. Virtual reality and 3D models ✓ | 7. Stringent and logical layout |
| 4. Predictive risk management | 4. Automatic failure handling ✓ | 7. No red zones ✓ | |
| | 4. Real-time monitoring and control ✗ | 7. Plant for showcasing ✓ | |
| | 5. Four independent hygienic barriers ✓ | | |

**Figure 9** Final list of innovation for the Tisso II project. (*Source*: Kalundborg Utility)

## Final result

In the end an innovation catalogue was prepared, see Figure 9. The catalogue included 21 innovations that were agreed to be included in the project. This is a good result considering that most of the innovative ideas were grounded in the needs of key stakeholders and understood and deemed relevant by all participants.

Unfortunately, at a later stage, the project had to make severe budget cuts due to the higher than expected bids for the building part. This meant that six innovations were taken out of the project. Three of them could be included again later (marked with an exclamation mark), others (marked with an 'x') required more substantial changes.

Still, it is a plant with 15 innovations. The most important being the ability in full-scale to supply water of drinking water quality from surface water without the use of chlorine, an innovation directly linked to the headline 'respect for water'.

**Your reflections:** What does it mean to take an innovative risk? How can collaboration attenuate innovative risk? How do you get inspired from the outside? Is there a culture of personal leadership and taking personal responsibility in the culture you work in? How does that influence your willingness and ability to take responsibility and show leadership? What barriers do you see for showing personal leadership?

## SMALL-SCALE IWRM IN LAKE TISSO

Lake Tisso near Kalundborg serves many purposes and poses threats for many stakeholders: Water is abstracted for water use for several industrial usages using different water quality. The lake, with its upstream and downstream rivers, makes up an important diverse eco-system. Cattle are grazing at the shores of the lake and the rivers. Commercial and recreational fishermen are fishing. Many people are using the area for recreational purposes, including sailing. The area is an active resort for tourists. Bird watchers are looking out for rare migrating birds, special designated islands in the lake need to be above water in the important hatching period and flooded clean in the winter. And lots of people live around the lake and its streams either permanently or in summer residences.

A principal point in the lake on which all of these activities depend is an automatic weir that controls the water level in the lake. The weir is currently controlled by the local municipality based on a 20-year-old regulation scheme. Even in this modern age, such an aquatic system with lakes and up- and downstream streams constitutes a kind of unrecognised centre for many people. This can be observed by the long-time criticism that has been taking place by more or less all of the stakeholders. In the latest municipal election, the lake control had become an important theme in the election debate.

As the water abstraction permit was to be renewed, the utility and the municipality decided to join forces to see if it was possible at the same time to improve the whole water system by looking into the weir system and its control. This decision was the beginning of an engaging Integrated Water Resource Management (IWRM) project. By including representatives for all stakeholders, this technical project has become an interesting Integrated Water Resource Management project. Representatives from all stakeholders were summoned in a process with the purpose of developing a new control strategy for the automatic weir that if possible would meet everybody's needs or make a fair and transparent compromise. At the same time the strategy had to take into account the expected future changes in precipitation patterns due to climate change with wetter winters and drier summers.

To aid the process two dynamic models are set-up for the area. A small local model simulates the lake and a larger holistic model includes the key rivers and the influence of the nearby ocean water level. The project takes place in close cooperation between the local water utility Kalundborg Utility and the local authority Kalundborg Municipality. The project has a high level of political attention as the lake is perceived as a major local asset and an important water resource for the Kalundborg Symbiosis, a thriving industrial ecosystem famous for its longstanding exchange of resources between different industrial entities.

The project addresses several SDGs with a special focus on SDG 6 (Sustainable management of water). According to the UN, putting Integrated Water Resource Management (IWRM) into practice will be the most comprehensive step towards achieving SDG 6. This small scale experience has taught us that IWRM is a very rewarding process in the sense that it provides a shared holistic and detailed understanding. This provided the opportunity to make a really innovative and effective solution taking all needs into account. At the same time it is a complex task that requires pro-active planning, deep technical-scientific understanding and a strong capability to mediate conflicting needs.

The project also addresses SDG 15 (Terrestrial ecosystems) by taking special care in adjusting the system to ensure the corresponding required conditions to enable a thriving ecosystem in the lake system. A key focus is on the indicator fish species trout. According to local investigations carried out by the local fishermen, the number of trout in the system has been falling drastically during the last decade. Additionally, there is a focus on vegetation, where a local botanic organisation has mapped the number of different species of plants for the area to more than 600 different species. One success criterion for the project is to establish a continued collaboration with local nature organisations to enable continued improvement after the project of establishing a new weir system.

In the centre of the IWRM project we established the 'Lake Tisso Forum' consisting of representatives from the NGOs, the industry, representatives of agriculture, landowners, summer residents, the river guilds, leisure organizations, the symbiosis centre, the utility and the municipality.

The forum process was planned with four workshops over half a year but eventually ended up including five workshops before a final agreement was reached. As in the case of the Tisso II advisory board process, there was again this from-workshop-to-workshop planning method. The results from the workshops were in a kind of dialogue with the modelling work, where every workshop was in some way a response to the result of the former workshop and each new modelling result was a response to questions posed at the previous workshop.

The establishment of a hydraulic model for the whole hydrological system, see Figure 10, was a central learning tool for both the project team and the forum. Besides the lake itself, it included the upstream river (Upper Halleby River) and the downstream river leading to the sea (Lower Halleby river) as well as a major side stream joining the Lower Halleby River (Boestrup River). The surface of Tisso lake is 12.3 km$^2$ and up to 14 m deep. It consists of freshwater and is located in a Natura 2000 area within the area of Kalundborg Municipality. Input to the simulation model include historical rainfall, water heights, water flow in Lower Halleby River, cross-section measurements of the river profile, water levels in the Great Belt, the sea where the river has its outlet through a delta called 'the Bottle' as well as through an artificial canal called 'the Sugar Canal' originally serving as an outlet for an inland sugar factory, that does not operate anymore.

The model served three different purposes: (1) as a basis to understand the system dynamics; (2) a system to compare different situations; (3) a system to develop and

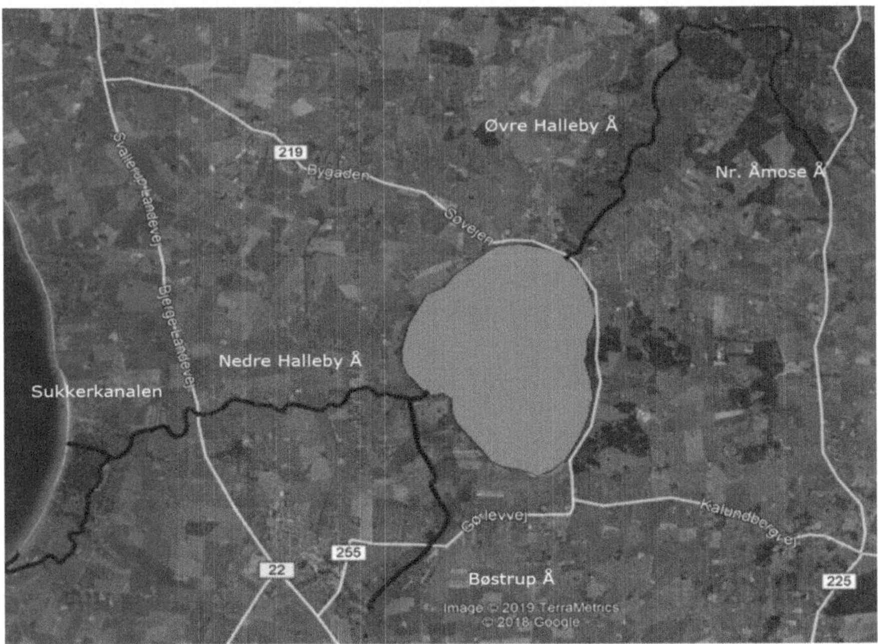



**Figure 10** The hydrological system of Tisso. (*Source*: Google maps)

compare newly developed control schemes like a benchmark system; and (4) a system for understanding extreme events due to climate change. But first and foremost it worked as a 'myth buster'. We, as humans with our human brains, do have an extremely poor ability to understand dynamic water systems in our minds. Hence, if we are not supported by models our conclusions are often plain wrong. This was very clear at the beginning of the project. We were all trying to comprehend and imagine how the dynamics work. A model is essential to gather people around a shared science-based understanding (almost like a microcosm mirror of the discussion of atmospheric $CO_2$ and climate change).

On a superficial level, this may appear to be 'an easy project', where one just has a dialogue with the stakeholders, provides simulation results, and develops a new method together whereupon everybody leaves the work happily. Not so easy, though. An obstacle at times seemed to be the mere fact that many thought this was easy. But environmental decisions are often only 'easy' when looking on the surface.

The challenges included bringing the very different stakeholders with their generally low level of understanding for water systems and hydraulic modelling as well as differing worldviews and interests along the whole process. Because of the initial widespread dissatisfaction among stakeholders, the process started in a state of negative trust capital. Hence, a first and continuous priority was building trust capital. Additionally, the hydraulic model understanding and mutual understanding of the many different stakeholder needs had to be increased dramatically both within the project group and among everybody in the forum. Remember, it is a painful process admitting 'not knowing'.

In parallel to running the forum process, the project group had to understand the hydraulic system themselves. They had to understand the system to the extent that they could continuously present preliminary results to the forum. The group needed to understand the controllability of the weir, the annual and seasonal variations in rainfall and evaporation; they had to understand how the operation affected the ecosystem and a number of other issues about the full system.

Hence the process was demanding from a technical-scientific point of view, from a facilitation and diplomacy point of view as well as from a pedagogical point of view.

## Workshop 1

Start-up of the forum. Throughout the forum a local politician acted as the moderator of the meetings. This was beneficial for a number of reasons. One crucial point was that the political system thereby gained an in-depth understanding of the results from the process so that they could later treat the recommendations politically in a competent manner. Another important point was that it provided the project team with a frame to step into, rather than also having to 'hold the space' of the process in the meetings. Additionally, the politician chose to comment on the project result stating his view, which supported the forum in its alignment process. This latter effect was really contingent upon the politicians' good abilities to make this come about. A different 'personality profile' may have had the unfortunate effect of doing the opposite …

The first workshop included an introduction to and a discussion of the purpose of the process. This allowed everybody, project group, politician and every stakeholder, to articulate their own hopes and understand the hopes of everybody else. All stakeholders had 2–5 minutes to present their 'case', their reason to be in the process and what kind of problem they hoped to have solved. The session was intense and interesting, and it underlined the diversity of interests, worldviews and personalities in the group of the 20 stakeholders.

Each stakeholders key-points were condensed into one or two sentences that were mirrored back in the referendum from the day. This was a mutually benefitting way to do it. The project group could make sure that they had understood the core of the messages. And, the stakeholders were assured that they had been heard and understood and that their interests were now being adequately considered. This was an initial action that started the process of establishing a secure base for all involved while increasing trust capital. Thereby, the feeling in the forum transformed from scepticism and suspicion that this was a 'pretend democracy process' to an increased sense of trust and engagement in finding mutual solutions.

This trust-building progressed differently in each individual. At some point in the process, the most sceptical participant challenged the group to come and see reality together with him. He did not feel that he was being understood deeply enough. So we did. It all contributed to the forum moving along despite conflicting interests and different worldviews.

## Workshop 2

Everybody involved in the process had differing understandings of the system dynamics and the implications of the natural part of the system, i.e. the effects of weather and physical dimensions of the system. Those who understood it in one way could not convince those who understood it in a different way. Did the problem relate to the lake? Or the upstream or downstream rivers? And where and when exactly was the conflict? And what in that conflict could be affected by the weir? And what were the consequences of changing weather conditions from year to year?

A brainstorming workshop where the stakeholders were mixed in groups of different and opposing interest was initiated. The purpose was for each group to identify and clarify the potential and real conflicts in each of the three parts of the system: Lake Tisso, and the Upper and the Lower Halleby River. This had to be considered during three different conditions: (1) high water level conditions (flooding); (2) standard conditions; and (3) drought or low water level conditions. Based on the groups' collective insight they were to come up with collectively developed solutions and ideas for solving or alleviating the conflicts.

The effect was that a lot of ideas and interests were unearthed from the silent tacit knowledge in each stakeholder. Additionally, the groups had a lot of fun with each other and the process included both laughter and eureka moments. This helped further loosening the tension and building up trust. Hence, the groups' presentation rounds were carried out with good humour and feel-good emotions in the groups as well as a play-pretend competition between groups about whose ideas were the best. Therefore, everybody left the workshop happy, hopeful and full of inspiration and new ideas.

However, the sheer number of ideas posed an unforeseen dilemma in the workgroup. The group had nowhere near sufficient resources to analyse all of the ideas. Some of the ideas were so far removed from the focus of controlling the weir that they were entirely out of scope for the project. This was, of course, a positive dilemma, but nevertheless a dilemma. In such a case ensuring that the process does not become a 'fake democratic process' was not so easy suddenly.

Finally, what we did was to prepare an idea catalogue and parked this in a 'track 2' of the process. The local politicians would then later have to prioritise these ideas after the weir control project was finalised. We made sure the ideas were described well enough to be recognisable for all involved. Some ideas were the combination of input from more groups, so the outcome did not end as one group or another 'winning'. Rather, in the end, there were 10 integration-ideas to work on later. This approach was easily accepted by the forum.

## Workshop 3

At this point in the process the modellers had finalised the setting up of the model for the hydrological system. The model, therefore, was presented for the forum. The stakeholders hardly had any experience with hydrological models – or models at all. So the presentation was to provide this understanding to the extent that the model could be useful for everybody. This invariably included understanding model assumptions, model calibration and model uncertainty. The main aim was to communicate how and to which extent one could trust the model. To explain that even if the model had some deviations from measurements, it would be possible to compare different strategies to understand overall patterns and success parameters. This was difficult material for everybody and spurred minor controversies between some stakeholders and the modellers. However, the presenter had a high level of authenticity, in-depth knowledge and pedagogical skills and some members of the forum understood the work and could hence help some of the other members to get the main points. Hence, generally, in the end, most members went home from the workshop with at least some modelling understanding.

Additionally, the forum was shown the Tisso II plant to get a more tangible understanding of the scale of what we were working with. The plant is quite impressive to see with all the pipework, the reactors, sensors, controls etc. So this also had an effect of improving trust in the capabilities of the utility and an understanding of the 'professionalism' and the 'industrialism' that was also part of the story about the lake. With its modern design, the plant radiated a different more impressive image than the more old-fashioned worn-down wood and concrete weir system at the outlet of the lake could.

## Workshop 4

This fourth workshop was held in buildings near the weir system. Hence a 500 m walk was arranged for all members to come down to the lake and see the machinery and enjoy the view over the impressive lake. The workshop was held in a nice museum focused on the area around Tisso, including material from the Viking age. This made a comfortable setting for the workshop. In general, each workshop was held in different places in the municipality to show the members different things and provide various kinds of ambience in the workshops.

In this workshop, the initial control experiments were presented and discussed. Improvements were suggested, and dissatisfaction was vented. In spite of the efforts in the last workshop to provide trust in the model, some of the stakeholders distrusted the model. This was partly because the model produced insights that went contrary to the stakeholder's internal understanding. Different viewpoints were vented, and disagreements resurfaced. The moderator did an excellent job to keep the situation under control in a good sense. The difficult issues were about understanding the result, about changing internal stories and about the discrepancy between model and reality. To be able to continue in an ordered way, the workgroup suggested carrying out individual dialogues with smaller groups of stakeholders in order to discuss the model in detail and to deepen the understanding of the stakeholder needs. This was acceptable to all participants, and a series of dialogues followed the workshop.

Different illustrative figures were developed to convey the results effectively. A rich and practical illustration showed each year superimposed on each other, see Figure 11.

The picture made it clear for everybody how varied the years are due to varying weather conditions, and how far and how often the states of flow and water level deviate from the current steering curve. The stakeholders representing fishing brought an external expert from a university. We learned a great deal from this expert about how we could improve the system to ensure better conditions for ecological health in the system. The agricultural representatives gave us a lesson in agriculture and how diverse the needs are between the different farmers – this led us to understand how difficult the job as stakeholder representative was. We learned something new from each meeting and the questions and doubts of each stakeholder were treated seriously and the insights were integrated into the solution to as much extent as possible.

Following these one-on-one dialogues and the input we gained, the workgroup put everything together into a new integrated control scheme. This again was a

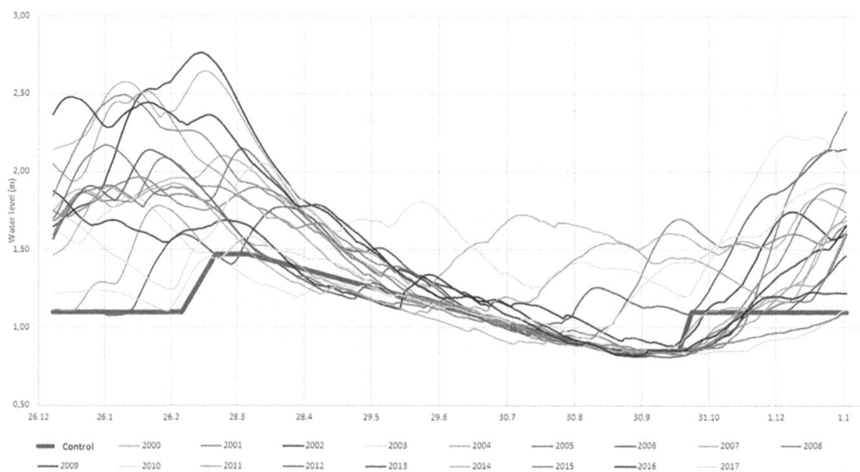

**Figure 11** Simulations of all years superimposed on one year. (*Source*: Kalundborg Utility)

complex process of different dimensions. It was challenging to find something that would work by performing well and at the same time, clearly signalled how each stakeholder had been included and their needs integrated – while at the same time not changing things too much. However, at some point, the core principles were clear, and the group reached the point where the new control scheme made reasonable sense from all perspectives at once.

## Workshop 5

Basically what the workgroup, the forum, the politicians and the steering group had learned during the process was the following four key learning points:

(1) The weir has little controllability regarding high water levels (flooding) – the low slope towards the sea is the main factor controlling the situation when a lot of water has to go through the river system.

(2) The water flow in the downstream river is the most critical element determining the thriving of the eco-system, especially trout. Sufficient flows at the right time are essential for the trout's ability to navigate the river and for adequate amounts of oxygen in the water. If the rivers cannot meet these conditions in most years, trout cannot thrive in the system and will eventually disappear.

(3) The current control regime (a steering curve) signals expectations that cannot be met. Hoping the system will follow this pre-determined water level curve has and will, if kept, continue to disappoint and frustrate everybody. Additionally, the flow in Lower Halleby River is essential and currently not controlled at all – with detrimental effects on the ecosystem. Hence, we must change our perception and expectation of the control scheme.

(4) The primary function of the weir is to ensure a proper withholding of water from spring to late summer to provide both a sufficient water level to supply Lower Halleby River to ensure the health of the ecosystem and to supply water to the industries. Both needs can be met simultaneously in non-extreme years.

These four learning points were not apparent from the beginning and each had in its own way and degree been subject to controversy. Now, these were settled and all agreed on these conclusions.

The new control scheme was developed based on these key insights to ensure a transparent and just compromise between all stakeholders. The new scheme enabled a significant improvement of the conditions for the ecosystem, while only reducing supply robustness and agricultural interests slightly. The results can be seen in Figure 12. As can be seen, the effects were benchmarked against simulations of the current 'steering curve'-based situation and a 'natural system' without a weir installed.

| Scenario | | | Supply | Landowners | | Ecosystem | | |
|---|---|---|---|---|---|---|---|---|
| Climate | Control | Abstraction | Supply not met | Max. Height surpassed | Average water levevl april-september [meter] | Qmin < 0.5 m3/s All year | Qmin < 0.7 m3/s Sep-Jan | Qmin < 1.5 m/s Sep-Jan |
| Existing | Current | 3,5 | 0% | 43% | 1,28 | 11% | 17% | 35% |
| Existing | W/o weirs | 3,5 | 12% | 41% | 1,22 | 1% | 18% | 48% |
| Existing | New | 3,5 | 0% | 46% | 1,40 | 0% | 0% | 17% |
| Existing | New | 5,0 | 1% | 45% | 1,39 | 0% | 1% | 17% |
| Existing | New | 7,0 | 2% | 43% | 1,37 | 0% | 2% | 18% |
| Climate 2050 | New | 3,5 | 0% | 45% | 1,41 | 0% | 0% | 18% |
| Climate 2050 | New | 5,0 | 2% | 44% | 1,40 | 0% | 1% | 18% |
| Climate 2050 | New | 7,0 | 3% | 44% | 1,38 | 0% | 3% | 19% |

**Figure 12** Performance in various control scenarios. The performance indicators in rows 'Supply', 'Landowners' and 'Ecosystem' in percentages describe the time where criteria are not met. The average water level is interesting for the landowners as it determines which part of the land can be accessed by machinery during sowing and harvesting. (*Source*: Kalundborg Utility)

The results show that the robustness of supply is unchanged, while the conditions for the ecosystem is improved tremendously – actually improved beyond the 'natural system' without a weir. The situation for the landowners is slightly worsened. An alternative control scheme was tested to improve this latter effect. That reduced this disadvantage considerably, though it was not possible to eliminate it completely.

The control scheme and its effects were presented to the forum. Two alternative control schemes were present and one was recommended and accepted by all in the forum. There was an almost audible relief in the forum about the solution, many saluted the result, and there was no critique except a few suggestions for detailed adjustments to the presentation.

Additionally, the effects of forecasted climate change were analysed. This had some implications for the industrial users, as the resilience tests of the system in regard to droughts showed potential problems with supply during late summer. This topic would have to be dealt with in a different project (see next case).

## What was upgraded?

This was regarded as a rewarding process as well as a satisfying outcome by all participants, which was a positive surprise when considering the starting point. In the worst case the result could have been a deepening of conflicts. A key insight was that the idea of secure base leadership works and it was fortunate that the project group was able to 'generate' enough trust to overcome the conflicts. At the same time, it is also apparent that it is a resource-demanding task to work this way with a large group of stakeholders. However, if – as in this case – the process eventually succeeds in creating peace and agreement, the municipality and utility will save many resources later due to the discontinuation of the more than 10-year-old conflicts. In the end, it may have been the least resource consuming way of doing it.

The demands on the skill set required to make such a process work is quite high. Especially, the three competencies of modelling and control, mediation/facilitation/forum leadership and political-diplomatic skills were all crucial for the success of the process. Had any of these lacked in the project or failed for some reason, the process would have been significantly less successful.

There is also an insight that a few setbacks in terms of increasing conflict level in the process can actually be a good thing. It focuses attention and helps everybody gain a deeper understanding of differing points of view; points of view that could otherwise have been unintentionally neglected or felt neglected by individual stakeholders. The conflicts and the way they were handled drew people together rather than apart.

This was a resource-intensive process for the workgroup – and hence costly for the municipality and utility. It is an eye-opener to understand what level of effort is required to carry out IWRM successfully. It really stresses the size of the challenge we have in front of us to transit to water stewards able to handle IWRM. The transformation to a sustainable agenda is a huge task. This cannot be done with a casual attitude; it requires attention, competence and respectful dialogue. But I believe if it can work out in this small-scale, it can also be done on a larger scale. But we really need to practice.

The use of models for this kind of problems is crucial. It is a pre-requisite for having a real and lasting effect. Even with a model, it is difficult to comprehend all the connections and patterns. Without it, the discussion is prone to myths and misinterpretations. Perhaps it would have been possible to reach an agreement without models, but that would entail a great risk, that the agreement was not built on the right system dynamics. The agreement would as Oren Lyons call it is not according to 'natural law' – which would become apparent five or 10 years down the road with recurring conflicts and without the hoped-for improvements for the ecological system. Hence, models are a vital tool for water stewards in many projects.

**Your reflections:** How well can you handle situations where you as the expert need to collaborate on an equal level with non-experts? Can you remain honest and humble? Can you navigate inspiring trust and confidence while at the same time be full of doubt and deliberations? Can you remain calm and friendly when your results are attacked? Can you articulate the needs of all your stakeholders – do you understand why they find these needs important? Can you find solutions that satisfy all needs? When you need to disappoint some stakeholders are you able to do that respectfully? What is your experience in building trust capital? Are you conscious of your methods?

# A MAJOR RENEWAL OF THE WATER PRODUCTION INFRASTRUCTURE

The infrastructure of water production utilities is often constant for decades before an opportunity for change arises. In Kalundborg Utility one of these opportunities has come up as the 70-year-old existing water treatment plant requires a major overhaul or, alternatively and more probable, be substituted with a new plant.

Besides its high age, the local industries predict an extension of production and hence predict an increase in water consumption in the near future. This further strengthens the incentive to renew the water production infrastructure. The renewal also provides an opportunity to establish a back-up solution for Tisso II for drought situations. Additionally, it provides a cost-effective way of fulfilling a municipal request of always having 1 Mio $m^3$/year in reserve in case new industries were to be established in the municipality.

The challenge is made further complex due to the Industrial Symbiosis in Kalundborg. Kalundborg Symbiosis is the world's oldest and most comprehensive industrial symbiosis. It commenced in 1961 and has developed progressively step by step since then. The Industrial Symbiosis now consists of nine private and public companies exchanging water, energy and materials through 22 different streams. Delegations from all over the world visit the industrial symbiosis in Kalundborg every week of the year. The central principle is that waste from one company becomes a resource at the next company.

The cooperation between the companies in the symbiosis provides mutual benefits, economical as well as environmental. The Symbiosis creates growth in the local area and supports the companies' environmental and climate change mitigation efforts. The annual gains of the Industrial Symbiosis for the partners include 24 Mio Euro indirect economic savings, 14 Mio Euro in socio-economic savings, reduction of $CO_2$ emissions by 635,000 tons per year, saving of 3.6 Mio $m^3$ water, saving of 100 GWh energy and the reduction of consumption of 87,000 tons of material. This of course morally and practically obliges the new water structure to show foresight in regards to identifying and realizing additional symbiosis and sustainability opportunities.

The default base scenario for handling this new situation is as most utilities would do: build a new water treatment plant and establish equipment to abstract the additional water resources – in this case, by establishing two additional groundwater fields.

While we are keeping momentum with the base case scenario by identifying new locations for the plant and groundwater extraction, we are at the same time searching for innovative alternative concepts. In the age of sustainability, SDGs, circular economy, new water treatment technologies, industry 4.0 and in the city of the world's oldest and largest industrial symbiosis, it is imperative to succeed in changing the water supply system markedly.

Hence, an innovative parallel track of the project is established. This track is a collaborative process to explore alternatives. The design of the innovative co-creative process builds upon the experience from the above-mentioned projects. As input to this process we used the idea of 'upgrading'; by asking: 'how can we further upgrade the innovation process to ensure that the group of people come up with a solution that is closer to a sustainable ideal, while at the same respecting what came before and provide solutions that are acceptable to all direct stakeholders?'

In the former case stories I have described how the projects 'upgraded' our competence, methods and mindsets as we went along. What we want to do in this project is to plan a series of project upgrades each born out of 'a frustration' with something that did not work or just did not work out right in earlier projects/experiments. The upgrades do not necessarily build on 'complete failures', but may also address a potential that did not bear sufficient fruit, a way of thinking that was not explored enough or something that was not fully comprehended or included. Working with frustrations and upgrades is a way forward, where we attempt to bridge the gap between what we are and what we want to become serious – and then try to do something practical about it.

At the stage of writing this, we are still in the start phase of the project. So what follows is a description of a number of 'upgrades' we plan to work into the project.

## Upgrade 1: Consciously attempt to express an ambition level for the result

*Frustration*

When working in a project group that works with stakeholders there are almost always one, or a few, who change each subject into a discussion about money too soon. This often means a derailment of the discussion. People who were supposed to imagine solutions start 'throwing things off the ship' too early with the argument that they are not absolutely crucial for the project and will entail additional cost. This causes a non-constructive discussion and means need to be taken to divert the discussion of cost until a later stage.

*Upgrade attempt*

For the project, a manifest of ambition has been developed collaboratively between the director and project manager. The manifest includes 15 challenges to the project; some examples are:

(1)   The aim is to create a 'next generation' solution for the water infrastructure, corresponding to keywords like holistic, new process technology, water stewardship, symbiosis and industry 4.0;
(2)   Production price should be no higher than today (as a 10-year mean);

(3)  Quality of machinery and buildings should be optimised according to the principle of 'total cost of ownership'.

(4)  The work environment on the plant should correspond to the principles of Vision Zero (a national initiative towards the design of workplaces with the aim of having no work accidents).

It can be very helpful to clarify and define ambitions for a project as illusrtrated above because such statements help the group manage expectations and reconcile conflicts over values. There may still be disagreement on the interpretation of the challenges, but there is a framework for this discussion which will hopefully prevent the discussion from becoming arbitrary. Then comes the discussion of cost and optimisation.

## Upgrade 2: Voicing expectations for the project execution

### *Frustration*

Sometimes a project group gives up too early on 'hard problems', they tend to 'demand a decision' from outside, e.g. from top management. Often top management don't know the answer either and instead, they want the group to help guide them towards a proper solution. Since many of these challenges are 'wicked problems' it is even more crucial that the working group can cover many aspects of the problems and possible solutions, and surely this includes much more than technology. It includes people, competences, organization, public relations etc.

### *Upgrade attempt*

Hard questions are questions that are important issues where no answer is available from the start. Instead, top management and project management collaboratively establish ground rules for how to approach a solution in dialogue. Hence a series of CEO challenges to the project team has been declared, some examples are:

(1)  ***Project economy***: establish continuous economic overview and transparency. When the economy is only vaguely defined, try to work with intervals of cost for the key factors. Then the cost estimates will become less uncertain as the project progresses.

(2)  ***Social sustainability***: Work with at least one element of social responsibility.

(3)  ***Water quality requirements***: The project should facilitate the decision on water quality requirements in regards to, for example, water softening and the treatment for pesticides between stakeholders.

This is both challenging and helpful. For example, in the first declaration about project economy, it would have been more convenient for the project group had it just been provided with a max sum. But obviously, a max sum in this case depends on many unclarified issues, which makes it impossible to provide at this stage. So instead the call is for transparency and overview of and quantification of uncertainty. Based on such an analysis it is possible to have a constructive debate. In declaration 2 for example, it is clear that that there is a requirement for an element of social responsibility in the project, but the way this is to be solved is not imposed on the group. The project group is in better shape to find a meaningful way to integrate the element in the project as the project develops. In declaration 3, again, the solution has to be negotiated with the stakeholders taking both requirements, wishes and cost into consideration. The CEO does not know what the customers and other stakeholders are willing to pay for different options at this point. The project must establish a process to find that out.

## Upgrade 3: Setting goals for sustainability
*Frustration*

Traditionally, the main balance in water projects is between quality/quantity and project costs. Sustainability often comes into the discussion as a 'side remark' or as something taken care of by legislation requirements from the local authorities' terms of permits. This, as we have seen, is a too shallow approach for substantial, sustainable change to happen.

*Upgrade attempt*

In the framing phase of the project model, goals have been set for sustainability. The goals are more aspirational than numerical at this point due to lack of knowledge, so eventually they need to be translated into 'smart' goals as the project progresses. The defined sustainability goals or aspirations are defined mostly in terms of applicable frameworks of working with sustainability:

(1)  All-new infrastructure elements have to be prepared for water stewardship certification according to the standard defined by the Alliance for Water Stewardship. Whether it is to going to go through the certification process is 'to be decided' AWS (2019a, b).

(2)  Calculate and minimise the water and $CO_2$ footprint (Water Foot Print, 2019).

(3)  The building should live up to the DGNB standard for sustainable constructions. If it is to be certified is 'to be decided' (DGNB is the German Sustainable Building Council's standard for sustainable buildings. Denmark has to a large extent decided to use that as the primary standard as well (DGNB, 2019)).

(4)  The infrastructure should aim at being effectively climate-neutral no later than 2050 as per the national aims for carbon neutrality; emission should

be reduced by 70% before 2030 (Climate Home News, 2019). As the duration of the project is shorter than this horizon, the project must instead find a technical solution that can be adapted in the future to meet the goal.

(5)   Innovation initiatives should include at least one new element of symbiosis and circular economy.

Certification systems are practical tools to effectively get around the critical issues of sustainability. The systematic analysis and goal setting in these certification standards increases factual understanding of the sustainability challenge, differentiates and raises the bar for the defined goals and ensures a higher level of commitment organisationally and individually.

However, there is always the bureaucratic challenge of keeping the certification documentation up to date. Certification and engaging with bureaus of certification causes a workload that to many too often seems excessively administrative and hence diverts attention from more pressing issues. Kalundborg Utility is already certified according to ISO 9000 (quality), ISO 14000 (environment), ISO 22000 (HACCP) and ISO 45000 (Health and safety). There is a limited capacity as to how many certifications can be maintained at a reasonable quality and understanding.

I find the question of certifications – as suggested above – difficult. The rigour of such an investigation and the commitment to 'answer all questions' has a positive self-disciplining effect and ensures alignment in the project team or the organisation – ensuring a mutual understanding. But the administration and the paperwork are burdensome.

## Upgrade 4: Using silent breaks for focus and attention

*Frustration*

On a micro level there is often a lack of attention in meetings. While this is annoying on a day-to-day scale, it may also have long-term effects. The lack of concentrated focus and attention was a key challenge in the Tisso II project. Focus and attention are important just to get the job done. It is even more so if we want to succeed in doing something complicated, integrative and extra-ordinary. There has to be a sensibility and awareness that is not there by default. People often have too many tasks and projects on their minds and at their hands.

*Upgrade attempt*

In the group, the idea of using 2 minutes of silence before commencing a meeting was discussed. I expected the people in the project group would find it too weird, so it was suggested as an experiment. Everybody in the project, regardless of background, found it to be an interesting and worthwhile experiment.

The experiment was straightforward and without a long list of requirements. The only instruction was to sit quietly for 2 minutes after the agenda had been presented, not talking, not reading, just silence.

The first few times we did this together the effect was quite notable. There was no doubt that it caused a grounding in everybody. It also allowed everybody to clear their minds and adjust to being in the project set in contrast to wherever they just came from. It also gave a kind of a friendly and intimate mood in the group in a good way. This has been so successful that it is now also used when the project team conducts external meetings. It has become a kind of 'identity marker' for the project and the project team.

## Upgrade 5: Replacing one expert panel with multiple panels

*Frustration*

In the advisory board process of Tisso II the three professors took the role as 'experts', and the rest of us took the role as 'listeners to experts'. But what became apparent in the example when the operator presented was that he was an expert as well in his domain. None of us had decades of observing and maintaining a waterworks plant. Most of us hardly had a few hours of experience. So why should he not be recognised and paid the same respect as an expert?

*Upgrade attempt*

In the new advisory board process we design the process so that everybody becomes part of each of their advisory board group. This means that meetings consist of several three-person sized advisory board groups – all sitting in a circle:

- A customer advisory board group: includes both private and industrial customers. They are knowledgeable in what the customer requires and hopes for, understanding at least to some extent, the trade-offs between requirements and the total customer cost per $m^3$.
- A professor advisory board with three professors knowledgeable in geology, industry 4.0 and circular economy.
- A utility advisory board group with colleagues from other utilities with supplementary knowledge in the field as well as their specific culture of how to do things. Some of the people on the board will also bring recent experience from similar projects.
- An advisory board group with young professionals. They are professionals in the water area or adjacent fields. Their job is to represent their respective fields of expertise as well as to some extent be the voice of the next generation – our collective conscience.

- An advisory board group with operators with internal experience and opinions about the operation of our waterworks. Those are the ones that will operate the system in the future.
- An advisory board group of designers and engineers on the project.
- A steering group advisory board group consisting of the steering group.
- An economy advisory board group consisting of controllers and designers of the economic models for the system.

Having these respective advisory board groups ensures that each viewpoint is represented by a few people who can strengthen and align their input to contribute to the process. It also pays respect to those who would usually not be on an advisory board. And perhaps just as importantly, this construction relieves the pressure for those who would generally be the advisory board, i.e. the professors. They no longer need to 'know it all', but can more freely ask questions and come up with daring new ideas. Most importantly, the hope is that the ensuing dialogue between boards and individuals will bring something new to the table.

## Upgrade 6: Defining an emotional field to enter the project from

### *Frustration*

This is a difficult one. The frustration is that most people enter the process with the fundamental goal to take care of their individual needs, wishes and problems as they perceive them – which is fine. However, the problem is that no one person really truly takes the perspective of nature's concern. In the project about Lake Tisso, it ended up being the sport-fishing organisation who were the most outspoken advocates for nature – for their purpose of catching fish. Their concern spurred a solution that resulted in a significant upgrade for the environmental quality of the water system. But what if they had not been there? What if the representative had had other priorities? How can we assure that everybody balances their own needs with the need of nature?

### *Upgrade attempt*

In the Tisso II project 'Respect for water' became a vital headline and it had some good effect, but more could come of this. In this project the headline will be further extended and will be used more extensively throughout the project.

The new 'directives' are:

(1) Respect for water
(2) Gratitude towards water
(3) Humility in the relationship to water

We will attempt to instil this notion as an 'entering' appreciative stance in those working in the advisory boards. The hope is that it will help everybody to focus on water as the substance all lives depend on and defocus a bit on their own needs and the traditional 'resource-thinking'. At the end of each session, we will try to track back how this affected our dialogue.

## Upgrade 7: A different inclusion of operations

*Frustration*

Operators play a key role in the project as they will be the ones to operate the resulting plant, but even their importance increases if we want the system to be run according to higher-level aspirations. In the utility we have been struggling to bring all operators on board in previous innovation projects. This is partly due to the mismatch between the competencies of people executing projects and the competencies of people responsible for operations. One difference is that one group have a mindset of what we can create that will be finished years from now, the other group is more used to giving primary attention to the problem of today.

*Upgrade attempt*

Half the project group is coming from the operational department, so they have a substantial say in the project. Additionally, the chief of operations is included in the steering group. But what has become a central tool for inclusion is an operator's design manual. The project team has jointly developed the catchphrase: 'If things are not written down, they do not exist in this project' to help focus on documentation. This puts pressure on an important pain point for the operators, whose daily work is not focused on a lot of writing. They have a lot of tacit and silent knowledge that often does not arise until the plant is finished and they wanted valves to be placed differently or machinery to be operable from the floor etc. By visiting a large number of plants and noting likes and dislikes we have jointly written an operator's design manual, which currently consists of close to 200 observations of likes and dislikes. They came about based on photos we took at different plants. These observations will be reasonably easy to include in the future tender material and hence ensuring the preferences and ideas of the operators to be included. We hope this leads to a considerable improvement in operator satisfaction.

## Upgrade 8: Opening up to adjacent fields of knowledge

*Frustration*

How do we think differently when we are always the same people or people from the same tightly knit society?

*Upgrade attempt*

By inviting people with different skills into the advisory board process a small change is achieved. However, when this is a minority, they often end up accommodating to the mindset of the rest of the group. So, in order to strengthen that, they are supplemented with presentations from external matter experts. For example, the first expert visiting the advisory boards meeting will present an overview of the field of biomimicry as it relates to water. Biomimicry is an approach to innovation that seeks sustainable solutions to human challenges by emulating nature's time-tested patterns and strategies. The presentation will be the centre of discussion among the advisory boards. In this sense, a small seed of something differently minded is infused into the process.

## Upgrade 9: Consciously building up social capital

*Frustration*

Why do these large projects often end up being more conflict-ridden than happy achievements of great constructions?

### *Upgrade attempt*

In the Tisso II project we experimented with some radical methods to reduce conflict pressure on the project. It is difficult to evaluate if the project had been more conflict-ridden without these measures. Regardless, the conflict in the project was not very enjoyable. The pressure increased and decreased in strength, but there were only a few peaceful harmonious months in the five-year-long period.

A key hypothesis in this upgrade attempt is that social capital in the project makes the project more resilient to conflict. There are a number of reasons why this may be. First of all, it is more difficult to be unreasonably angry with people with whom you have a strong relationship. Second, more time spent together socialising may make it easier to pick up the phone and ask questions or raise concerns before issues become full-blown conflicts or failures that everybody wants to avoid taking responsibility for.

Hence it has been decided to engage in team-building activities. However, the group has opted out of traditional team building activities and have instead chosen to meet at each other's home every 2–3 months, where the host prepares food for the group.

## Upgrade 10: Creating maps of the holistic level

*Frustration*

Most people have a lot of opinions about the small stuff but tend to shy away from the top-level decisions of the project, where much more is at stake. Hence, holistic full-picture challenges are often left orphaned. While these are the most important questions, they are often also the most complex. How do we ensure that the advisory boards and the project group spend time understanding and coming to conclusions on these issues in a meaningful manner?

## *Upgrade attempt*

A significant difficulty is that the holistic level often remains abstract and invisible. So, in the upgrade attempt, we try to draw maps that explain how things are connected, thereby also helping everybody to gain a sense of proportions on the top level.

One example is drawing a map of the water distribution system with the current water flows and the flows as they are expected to change in the future horizon of 10 years see Figure 13. Such a map is useful for identifying options for a circular economy, for focusing on the right water streams and for determining capacities of the new system. For example, one water stream of an industrial facility may be easy to recycle, however, if it shows up on the map that its consumption is less than 2% of the sum total, it is not really helpful. So though it may be worth pursuing at some point, it is not a game-changer when speaking of different infrastructure concepts.

Another example of a 'map' shows the development in consumption over the last 20 years, together with a prediction for the coming 10 years (not shown). This plot gives an understanding of the uncertainty of the predictions and at the same time makes it obvious that building a system where the capacity can both be increased and decreased would be extremely favourable.

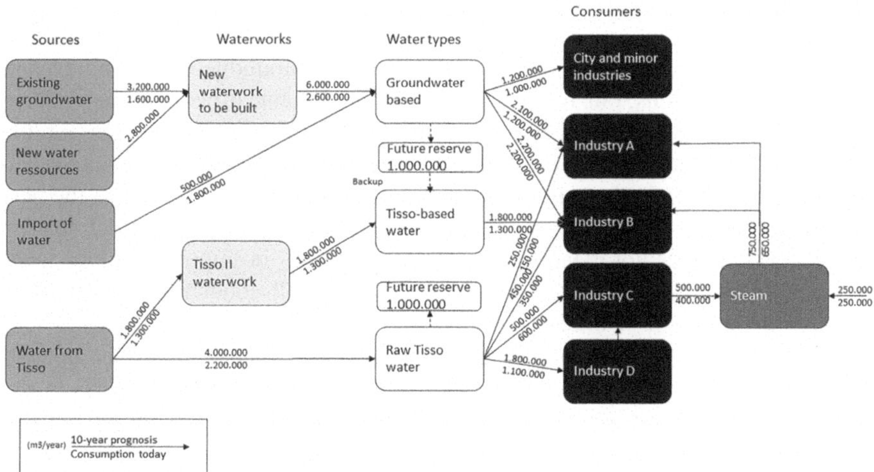


**Figure 13** Map of water streams. The numbers are confidential and hence fictional numbers are used here. (*Source*: Kalundborg Utility)

The project is still in its early phase and time will show which effect these upgrades will have, but just starting to map the high-level issues ensures focus and gives a sense that this will enhance learning.

> **Your reflections:** What frustrations have you experienced? Where have you been stuck? What actions could decrease the risk of that happening again? In what way does your work have long-term influence? How do you want that influence to play out? Who can help? How can you help others?

## REFLECTIONS

The inner sensing, studying and contemplation about water and water stewardship is in a continuous dialogue with what happens in 'life on the outside'. The projects in Kalundborg Utility spans six years and has been a 'field of experimentation' and expression of the new ideas. At the same time events, results and failures here have infused the contemplation process with fuel for new emotions and thoughts. Though it is difficult to reproduce the dialogue in writing, it is distinctly clear that such a conversation has taken place and continues to take place.

A personal theme during these years has been the inner work of building a stronger connection to the heart. This gave rise to trackable changes in discussions, in analysis and in decisions. When having a connection to the heart (as in practically having a conscious awareness of one's own physical heart), something physiologically happens that affects the brain processes. Things seem to slow down and give rise to a different kind of attention. In discussions of conflicting viewpoints, for example, it becomes profoundly clear that despite differences of opinions, our relationship is also important. It becomes possible to hold both – the regard for the feelings of the person and the concern for the overall health of the full system. When the emphasis on relationship is increased, something softens in both parties and common ground is easier found. In the best case, the differences in viewpoints become a strength instead of a conflict. On the other hand, this does not mean to avoid 'confrontation' in important topics with, for example, angry people or people of power. It is all about finding a balance with a grounded sense of integrity.

The heart is also helpful in ensuring self-care and self-protection. One need to consider one's own strength, capacity and balance together with the needs of the more extensive system one is embedded in – here the utility. Brené Brown formulated this balance quite well in 'we can't sacrifice the micro for the macro, or the macro for the micro' in the sense that we can't sacrifice our personal sense of harmony and equilibrium for the system's harmony and equilibrium, i.e. my own thriving and sense of balance for the results in projects. But the opposite is also true, i.e. I will not and should not stay in my comfort zone at all times. To find a proper method and rhythm for walking in and out of balance in a productive way has proven difficult. As I became more sensitive, I could observe how I repeatedly fell 'out of balance' in different ways during the projects and experiments. This process of losing and regaining balance seems today as a kind of underlying soundtrack to the conversational process.

Looking at this occasional painful process as a capacity or capability building process is helpful. Looking at it this way it is clear that integration in, for example, Integrated Water Resource Management, is not a one-off event. Instead it is a continuous integration process, where various aspects of the utility water cycle become increasingly integrated with its surrounding in the form of cities, industries and nature. And even that is too general; it is a process of steps of integrating with this or that part of nature, i.e. a lake, a stream or a groundwater basin, and it is a process of integrating with this industry and then that industry; and that part of town and then that part of town. Similarly, the road to the grand aim of sustainability or to reach the sustainability development goals consists of a multitude of steps taken in all organisations as well as in individuals' lives.

If this journey towards sustainability, SDGs or integration is viewed through the lens of 'a living process' it becomes clear that the purpose is not to reach a definite goal after which we are 'home safe and can rest from there on forever'. A living process continues forever. Hopefully, the world moves back into a safe, sustainable comfort zone again, but that will obviously not be 'the end of history'. The path ahead cannot only be about moving towards sustainability, it has to have another dimension as well – an aspect of 'truth, beauty and goodness'. What is clear of course is that the journey can only continue if we get into a zone of sustainability, but it cannot continue there if it is ruled by lies, ugliness and hardness. We crave beauty, grace and the poetic.

An insight or approach to the work that has helped to make this difficult process a better experience is that though it is 'difficult and often hard', it is also an opportunity to learn. It is well known that more is learned by failure than success. Thinking of this as a learning process takes a bit of the gravity off. A different way to understand this is that when we try to follow the heart, we may easily stretch the experiments so far that failure will arise at some point – only then have we stretched the ambition far enough. This provides a sense of forgiveness for the failures and avoids that we afterwards feel remorse that the bar should have been raised more. Hence perhaps we should stretch our goals so that we make small failures on a regular basis – not big failures and not no failures.

Working towards an ideal of more heart for water, i.e. better and more integrated management of utility use of water, more respect and care for water and nature, ensuring all relevant stakeholders needs through mediation, application of innovative technologies and better methods and codes of collaboration, has

**Figure 14** The Labyrinth of Chartres Cathedral. The process of moving towards sustainability and water stewardship is neither linear nor chaotic; but it is complex like the path through the labyrinth.

been and is an exciting process – it is a good and meaningful life. When going through the above examples, I can see how the process has overall moved towards this aim and how we have succeeded better and better. It is also clear that what we learned in one project is a stepping stone and a sounding board for the next project.

Still, there is a long way to a truly integrative, respectful, effective, nature nurturing system. We are after all still working within the limits of what is possible in many dimensions – technical and social. But boundaries are moving; people seem to become more aware and more caring in these times; not only here in the utility but also around us. The development around us is helpful; it makes it easier and more acceptable to move in this direction. At the same time, the above case stories are not unaffected by the polarisation around sustainability taking place on all scales globally, nationally and locally. Finding ways to understand the underlying needs of both poles is the key to finding workable solutions.

Hence the process towards sustainability is neither linear nor chaotic, it is somewhere in between. In Chartres Cathedral, a famous labyrinth is set into the floor stones in an area covering $130\,\mathrm{m}^2$, enabling you to walk through it. The labyrinth is not one of those where you get lost. Instead it leads you through a complex road that moves you in between the four main areas of the labyrinth. Though it generally moves you closer and closer to the centre, there are also times when it moves you further away. All the time moving you forth and back through overlaying stretches. I get the same sense when I try to think of the work and the journey towards water stewardship.

**Your reflections:** How would you describe the journey you are on – or the direction your work is moving? Is it visible in your work? How does it feel?

# *Chapter 3*
# A model for maturation

Our mindset is so difficult to see because it is ingrained in us – it almost *is* us. We cannot hold it out in front of us to have a thorough look at it. Rather we see the world through it. We may refer to it as 'common sense' because that is what it is to us. However, one notices that common sense is not the same to everybody. Ken Wilber compares the mindset to grammar in our native language. We can speak our native language and use grammar correctly without thinking about grammar. And it is not until the syntax has been pointed out to us in school that we would even suspect such a thing existed.

Human psychological development models are helpful in decomposing our mindsets. The model points out a pattern, where our mindsets develop in a way of layer upon layer. Such models have been developed largely independently by different researchers drawing models that are surprisingly alike. The words in the models are different, but there is no doubt that the researchers found very similar developmental paths regardless of their starting point. The benefits from these models are (1) it provides a clear and logical overview of the human development story stretching from the beginning of human life up until now, (2) the same steps that humans have taken on the grand evolutionary scale is taken by each individual as a process of growing up. When we understand that different ways of seeing the world belongs to different layers of this development, it becomes easier to understand 'the old story' and to some extent to navigate towards 'the new story'.

© IWA Publishing 2020. Water Stewardship
Author: Pernille Ingildsen
doi: 10.2166/9781789060331_0097

# GRAVES MODEL OF HUMAN DEVELOPMENT

Based on a lifetime of research, Professor Clare Graves proposed a stage-by-stage development model, which our historical mindsets/worldviews follow. The model contains a surprisingly few stages, only six to eight stages. Six stages have been included in our current mainstream world culture, while the two following stages belong to a second tier of stages, which is currently only available to a small but growing number of people. Graves believed that later even more stages would follow as the human cultural story progresses.

The progression of each new mindset stage emerged as a response to changes to life conditions and thereby caused an evolution of society and its capabilities, which again provided new life conditions, causing new changes to life conditions.

Graves' spiral model (Figure 15) of the evolution of mindsets starts with the beginning of the human species and follows human development through periods of hunter-gatherer cultures, agrarian cultures, the industrial era etc. Each era offered different life conditions for the people living in them. Hence, they adapted their pattern of thinking and acting accordingly in order to survive, adapt and thrive. This eventually created new life conditions to which the mindset again had to adapt, and in doing so it became like a step-by-step evolutionary process. When seeing the whole process from above, it looks as if the way it unfolded was almost deterministic, a process that was set in motion over tens of thousands of years ago and continued through its natural cause, stage-for-stage, pattern-for-pattern (Graves, 1971; Beck & Cowan, 2005).

In our own life it is no surprise that mindsets develop over time. What may come as a surprise is that it is possible to decipher in us the six to eight major historical stages on which the mindsets are founded. Each stage has their individual core qualities and values.

Historically, societies stay balanced at one level of mindset for a long time. The appearance of each new level emerges in a time of crises, where people and societies so to say 'outgrow' the old mindset. The new stage of mindset can be seen as an attempt to make effective changes based on the malfunctions of the previous level – like a child that grows out of one way of behaving to take on a more

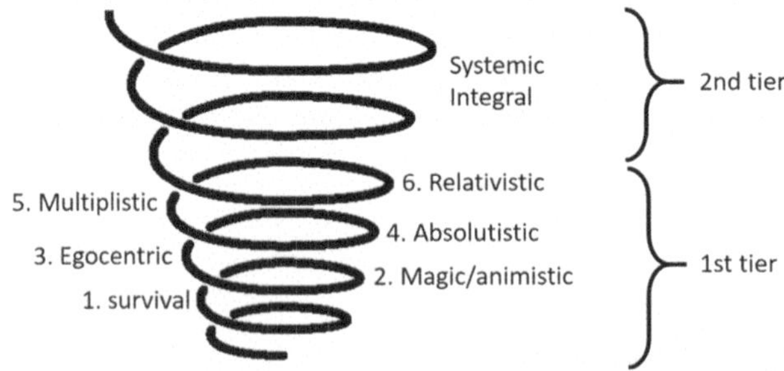


**Figure 15** Spiral dynamics model, Clare Graves. (*Source*: Adaptation based on on the work of Clare Graves)

developed behaviour. Between these stage changes, humanity thrives in relative stability.

Each new level has its negative aspects, its shadow side. As the shadow sides become increasingly dominant, the pressure for a change of stage increases. A crisis in a system can, therefore, be a precursor for a leap forward and upward. Alternatively, the crisis can lead to regression so that society moves back to a previous step. On a large time-scale societies can be seen moving forth and back between levels, before stable success in the progression is achieved. In order to succeed in overcoming the crisis by progression, societies have to move collectively or with enough 'mass' into a new more complex set of cultural values. This is not an easy operation. The times of such changes are tumultuous, and many momentous milestones in world history can be interpreted as struggles between different stages of development. Similarly, we can see a lot of the polarization taking place in the world today as an expression of different mindset levels grinding against each other.

The concept of value memes was proposed by Richard Dawkins to describe a kind of comprised cultural unit representing an idea (Dawkins, 1976). Dawkins saw the value meme as an analogy to the concept of a gene in biology. An example of a value meme could be the quote by Martin Luther King: 'I have a dream' (King, 1963). This quote comprises a comprehensive story and immediately invokes a sense of the values like equality and freedom. In this case, the quote related to equality of people of colour in the US. Another example could be 'the invisible hand' of Adam Smith (1759), which speaks of a different type of freedom, the individual's freedom to act in the market place without thought for the collective. The quote expresses that every man should freely pursue his interests because 'an invisible hand' will make sure that this, in the end, is in the best interest of the community as a whole. So while King's meme is about collective responsibility, Smith's meme is more concerned with individual autonomy. One way to comprehend and see a stage of the developmental process is to look at the belonging value memes. A stage can hence be characterised by a cluster of value memes that tell a story about the self-perceived role of humans.

The function of the value memes belonging to a mindset stage is to help everybody interpret and align themselves according to the values at that stage. Most individuals generally make this alignment with the unconscious purpose of belonging to his group of peers. The value memes work as a kind of mindset handrail to align people around these values. It provides a similar navigational map for everybody on that stage. This helps us to interpret each other and keep each other appropriately accountable according to the stage. Or in the case of change of stage, the new memes foretell and attempt to capture important aspects of the new stage, so that everybody can see it in order to move forward.

Graves identified six general stages that can be tracked in the process of the emergence of human beings up until today. As each stage occurs, the old mindset is transcended but, at the same time, there is a kind of integration of the old mindsets. The old mindset is carried with us on the journey ahead in a kind of 'inner' packaged form. The old ideas and memes still exist and create a kind of mixed amalgam mindset. Different mindsets can handle various problems so that problems are often handled only at the minimum required level of complexity. Depending on the developmental strength of the different stages, individuals may not have solid access to higher stages of mindsets and hence must handle their problems at the insight level that is available to them.

In that sense, the stage development is like a new 'shell' around the old mindset, providing new tools and different ways of looking at life and our role in life, similar to how Kalundborg Utility expressed their strategy as a continuation of old values but embraced by a new layer of values related to sustainability.

Ken Wilber (1996) who has studied the human developmental process, pointed out that Professor Graves' spiral model corresponds well to a number of other developmental models developed in other contexts. These are models from Gebser, Piaget, Kegan and Loevinger. Their models were developed more or less independently, but the stages that they identify describe a similar developmental process, though focusing on different aspects of the stages.

When we are looking for a vision for the water sector of the future, the models provide valuable insight. The description helps us understand how the general global human culture is moving, and by doing so, the understanding can both inspire and help us align. It can also point out some of the key features of the new story and help us transcend our current gridlocks. Finally, it can also identify how the value memes or key ideas in how we work today emerged, from which kind of thinking it came. Additionally, I think we can also see that our practice in water engineering has elements from all levels. Understanding the higher levels can provide inspiration for how to transcend and include current practices, peacefully and recognisably. Overall, I find this model to have interpretive value both for the future and for what is. A short description of the stages follows. Each stage is assigned a number and colour code as well as a few descriptive words. The reason for the colouring is that each of the stages is each so rich that the assigned words are only covering the patterns of thinking to a very small extent. To name it by a colour makes it more memorable and enables a continued expansion of the understanding of the facets of the mindset.

---

**Your reflections:** How would you identify aspects of your mindset that differ from others – not just difference in opinions?

---

## FIRST STAGE, BEIGE – SURVIVAL

The first base level is about survival. With this mindset we are occupied with our most basic needs of food, water, warmth, shelter etc. These needs are hard-wired into the human physique. Leaving any of these needs unmet for even a short period of time causes physical pain or eventually death, pains of hunger and thirst, but also a myriad of different pains and discomforts signalling that something needs to be taken care of in the body. For many generations we have had an infrastructure that is organised to handle all these needs at all times – at least in the more developed part of the world. Comparing to hunter-gatherer societies where survival depends on finding and hunting prey, most of us live in societies where these needs are met in an easier way. Today, in affluent countries, many people may not even experience thirst or hunger to any severe level throughout their life.

Water is one of the most fundamental needs of the body as the body consists of 60% water that needs to be cleaned and replenished. The water is used for a multitude of purposes in the body: to regulate body temperature, as a matrix for the production of a multitude of hormones and neurotransmitters, allows new body cells to grow, reproduce and survive, works as a shock absorber if the body falls or is hit and is crucial in the protection of the spine and brains, distributes oxygen throughout the body, helps the body stand erect, lubricates joints, flushes unwanted body waste in the form of urine out of the body, keeps membranes and helps digestion. Similarly, food is an important basic need as it supplies energy to the body, making it possible to move, keep warm, think etc.

It is rare to see grown-ups who only have access to this development stage, but it can be found in people who have severe brain injuries or Alzheimer's. According to the theory, we have all to grow up through the different developmental stages, and hence we all start at this square one as newborns.

In general, this stage is for obvious reasons assumed to be quite primitive. However, our forgetfulness of our body and its physiological needs may tell a different story; a story where in spite of this level being easy to satisfy, our 'body forgetfulness', bad drinking and eating habits contribute to unsustainability at this very core level. By having the basis of physical sustainability handled at an almost automatic level, we may have lost some of our own physical sense of the fragility of bodily sustainability.

Therefore, to get a more direct physical sense of sustainability is to focus attention on our own bodily sustainability. To sense how the water we drink and the food we eat sustains us. To make drinking and eating something we do mindfully. Some have taken drinking and eating to an art-form as for example, the sommelier who has trained the ability to recognise various taste elements in wine, chefs of advanced cooking is another example. Today, a few people work with water taste as an art – as water sommeliers. Some of the parameters a water sommelier rates are saltiness versus sweetness, smoothness versus complexity, contents of sodium, magnesium and calcium and the water's bitterness. The development of ability to tasting water is one example of how we can advance our appreciation of the most basic level.

In Danish, there is a wonderfully illustrative meme about this level which is often said when we have to go eat (it rhymes in Danish): 'Uden mad og drikke, duer helten ikke' – 'Without water and food, the hero doesn't work'.

> **Your reflections:** How much do you know about the taste of water? How can you feel that you yourself is more than half water? How does the water inside you feel? Can you remember being thirsty and having that need met with pure water?

## SECOND STAGE, PURPLE – MAGIC AND ANIMISTIC

The second developmental stage is where humans start organising survival in small collaborative groups, hunting, travelling and living together. 'You watch my back, I'll watch yours' would be an example of a meme belonging to the second level.

This level is called magic and animistic because the collective inner cultural world is full of stories of spirits. The surrounding world is alive with spirits and magic. Many cultures are still full of stories based on animistic understanding, where trees, water, rain, mountains etc. have each their benevolent or malignant spirits. People are working together with the spirits, trying to please them to make survival possible and life comfortable.

Many spirited stories are about water because water is still key to survival, not only of humans but also of the prey they needed for food. Veronica Strang (2004), a cultural anthropologist, explains that the water deities belonging to these early societies were often local beneficent creatures with which humans went into partnership with. That doesn't mean that they were soft, they could be dangerous as well and in, for example, Australia, water spirits were closely related to ancestral law and could swallow you if you transgressed. So the water creatures were powerful and could both generate life and take lives.

Also, this stage is not very prevalent in today's modern world but is most vivid in original cultures. When I feel – on occasions – that the relationship with water becomes too materialistic and 'dead' it may be that this level of my mindset senses something lacking.

> **Your reflections:** Are you attracted to the stories of myths? What may these stories bring that nothing else can provide? How is your connection to the myths of your culture?

## THIRD STAGE, RED – EGOCENTRIC

While the first and second stages are rare today, the third egocentric level is fairly mainstream. In this stage, humans work to gain a higher level of individualisation, trying to become 'one's own man'. The concept of 'power over others' comes into the arena.

The stage emerged when the groups from the second stage reached a level of collective infrastructure, where the pressure of external natural dangers had resided a bit to the background. The societies had succeeded in becoming so safe that strong individuals dared challenge the 'ghost stories' of the spirited view of things, suddenly perceiving the story as more imaginative than real. In the ears of the young, the stories stood in the way for their wish to be free and independent.

In the most benevolent forms, it caused a sense of being an independent individual, in its more malevolent forms it caused societies to be riddled with power games, wars and seeking of dominance. Extreme malevolent leaders working from this third stage include people like Hitler and Stalin, but people working from an egoic point of view is common everywhere. There is in this stage also power and radiance in people of this conviction, that make them look as potentially strong leaders for leadership positions. So they are found in leadership positions everywhere, often with negative outcomes for those below them.

'Power' can be defined as the drive or ability to affect change based on one's own will and motivation. If the will and motivation is positive, power is positive. Churchill was seen as a strong but also somewhat abrupt person, possibly because he had a lot of this stage in him. This was effective in handling the situation of World War 2, but it was not effective in the more complex world that followed thereafter. Hence, he lost the elections soon after the war.

In a water perspective, the water wars are relevant from this stage of operation. This is a stage that is best captured by memes such as 'survival of the fittest' and 'law of the Jungle'. Therefore the stronger and more fortunate, i.e. in this case upstream located communities, can dominate the less fortunate downstream communities by cutting off water. As this is intolerable for the downstream communities, war and power struggles ensue.

There are many examples of keeping water sources for oneself rather than collaborating in the world arena, i.e. between Israel and Palestine and around the water resources of the river Nile. But there are other more subtle ways of conflict, like neglecting to keep the waterways clean, in this way the downstream water is of a deteriorated quality and sometimes impossible to use for agricultural, industrial or domestic uses.

> **Your reflections:** How many of the people you work with are grounded primarily in this level of mindset? How do you experience that? Can you recognise the feeling of this mindset in yourself – in certain situations? Can you see both bright and dark sides of this mindset?

# FOURTH STAGE, BLUE – ABSOLUTISTIC

The fourth stage is organised around a 'law and order' mindset, where the impulsivity of people from the third stage is reined in by rules and regulation and hence the less fortunate in strength are again offered protection by society. Behaviour and interactions between people are regulated by laws or strong dogma. This is where large organised religions come into the world.

The story of King Arthur can be seen as a transition from a third stage society with warring lords and a country in turmoil to an ordered and peaceful country with a strong heartfelt dedication to King Arthur, a dedication that is not due to ruthless domination, but rather due to 'god-given authority' handed to Arthur by his drawing the sword out of the stone.

People are assigned clear positions in life, usually already from birth. Being the son of a shoemaker, you will live and die as a shoemaker. This is also the kind of thinking that can produce large empires. Empires could not be achieved solely with a third stage mindset, as the power is concentrated in one person at that stage – that is simply too primitive to base the foundation of an empire upon. The radius of the leader's power is limited to his personal outreach extended by his first-line henchmen. To build empires requires organised courageous men with hearts burning for a cause or a 'god-like' leader – and it requires rules and regulations.

It is a stage of great loyalty towards one higher god and one cause as exemplified by the crusades. Historically, at this stage 'the divine' transitions from the many spirits or demigods at stage one and two to the one God placed high up here at stage four. Margaret Strang (2004) explains that water spirits are portrayed as serpentines and dragons to be slain by heroic men in service to God above. Water spirits are no longer something to be partnered with; it is something to gain dominion over. Small amounts of water are made holy by the touch of the priesthood and used in holy transformations such as baptising. Hence, there is in this mindset a wish to control and dominate nature – and water.

Not all rules and regulations can be said to stem from this mindset, but the idea that rules can be used to rule the world and blind faith in these rules belong to a worldview based in this stage. I have good colleagues who, when asked of their work purpose, state that it is primarily to manage water so that they comply with the relevant regulation. Nothing beyond compliance is required, it is almost 'illegal' to do better as it may cause an additional cost. This is also the mindset that Christina Woods spoke out against in her paper 'You can't negotiate with a beetle' (2010) – interestingly supported by ideas with roots in an indigenous look at the world. The idea of 'natural law' as proposed by Oren Lyons (2004) is founded in first and second stage experiences and memes.

> **Your reflections:** 'You must, have to, shall, should' – how would a day without these words look like? If ordering yourself and others around was not an option, how would you get things done?

# FIFTH STAGE, ORANGE – RELATIVISTIC AND RATIONAL

At some point the fourth level society became perceived by some as over-regulated. It had a confining feel to it, and there was a need for individualisation – again as in stages three and one. The fifth level is a rebellion of the individual. The absolutistic system had become too clever in regulating every part of life and individuals longed for their personal freedom again, perhaps especially those who had been assigned less generous roles in society. However, it is a different kind of freedom than the ego-centric freedom of stage three. It is the freedom of, for example, Martin Luther (born in 1483), who insisted that people can think for themselves and apply their own good interpretations of the Bible, thereby undermining the power of the clerical priesthood in Germany and later in large parts of Europe.

He thereby provides the Christian God with a new place, not in churches but in each individual's own mind, heart and interpretation. This was revolutionary! Suddenly, the Bible became translated so that everyone could read and think for themselves. This is a clear break from having clergy interpretations of moral standards told from above by a priesthood that appeared at times to interpret the word more in line with power dynamics (stage 3) than as intended.

Stage five also marks the beginning of the age of science, invention and rationalism. The idea of God is challenged; the earth is (in our minds) moved from the centre of the solar system to a place less divine, orbiting the sun together with other planets. Rather than seemingly arbitrary God-given rules administered by the top of the hierarchy, the Church being too much affiliated with the 'worldly' power, individuals start thinking independently and their position on the societal ladder is no longer determined by birth. Social mobility is enabled by merit, industriousness and inventiveness. Adam Smith's concept of 'the invisible hand' (1759) belongs to this mindset. Hence it is a time for increasing competition as hierarchical positions become fluid rather than pre-defined from birth. The industrial revolution is rooted in this mindset where the harnessing of heat and water marks the onset with the invention of steam machines. Thoughts of economy governing the lives of individuals and society to the extent that brings us the crashes of Wall Street and the power of large corporations belong here. The underlying mindset is one of science and rationality. This is still the dominant mindset of this age.

However, it is also the stage where taking a third-person perspective grew as a mindset tool for everybody. Great political changes appear at this stage. Equality of gender and race is a key battle. Slavery is abolished in many countries for the first time in history. Women gain the right to vote.

At this stage, the human understanding of time changes from a grounding in the annual cycle of time defined by the changes in seasons to an understanding grounded in time as moving forward – time is stretched out before and behind us. Change, improvement, development, visions, to the ability to think of alternative scenarios and ways of organising is taken to a whole new level. This concept of time stands in strong contrast to previous mindsets anchored in a circular understanding of time where spring follows winter.

Water infrastructure advances notably under this mindset, from latrines to sewer systems, to wastewater treatment plants, to sensors, control and automation. This is the stage where Edward Snow found out the role of microorganisms in some diseases. There are no more spirits or gods to be found in water. Water is a cheap resource for the progression of human lives, and the focus is all on human comforts by engineering in the urban water cycle.

While each change of stage has led to a significant increase in the comfort of living for people, no single stage has been so transformative to earth and material capabilities as this change. The results of cutting-edge science within all areas of life must have been almost unimaginable at the onset of this stage shift. The human race is so successful in gaining resources for itself that it has outcompeted every other species for all relevant kinds of resources. Nature is no longer perceived as a threat and where nature inconveniences are experienced, technical solutions are found.

The shadow side of the fifth stage is global degradation and break-down of natural sustainability of major life ensuring systems. This is not to be misunderstood so that this stage is synonymous with unsustainability, but the means with which the human race can be unsustainable are so powerful at this stage that the global ecosystem is impacted, not as in early historical periods, where local areas could break down, and be left again to recover later.

> *'Men and women, everywhere and at all times, have despoiled the environment, mostly out of simple ignorance.*
>
> *Modernity's ignorance about the environment is much more serious, simply because modernity has many more powerful means to destroy the environment.*
>
> *Tribal ignorance, on the other hand, was usually milder; but ignorance is ignorance, and is certainly nothing to emulate.'*
>
> Ken Wilber (1996)

---

**Your reflections:** What are the bright and the dark sides of this mindset? If you were to move forward and hence both had to transcend and include this mindset; how would you do that? What would you bring with you?

---

## SIXTH STAGE, GREEN – RELATIVISTIC

This stage is a reaction to the loneliness and dispiritedness of the fifth stage. At this stage the primary focus is on harmony, everybody being heard and everybody being equal. This is a reaction to the raging competitive individualisation at the previous stage; this is the emergence of postmodernism. The sixth stage is green, the counter flow entering the public world stage in the flower power movement of 1968 in the western world. Later this results in new types of companies like ecological food markets. The mindset tries to bring an egalitarian worldview to the centre stage, despite all difficulties. The mindset is most easy to distinguish in the Nordic countries and in cities like San Francisco. The Martin Luther King quote of 'I have a dream' belong to this mindset; it is the stage of Occupy Wall Street, Greenpeace, Amnesty International and Doctors Without Borders.

The sixth stage arrived in the water sector with a turn of focus towards the environment. Due to the water sector's dependence on one of the most basic elements in nature, the water sector had to understand and deal with environmental concerns early. As it became apparent that the industrial handling of water had severe ecological effects, the focus in the last quarter of the 20th century has increasingly been concerned with the state of the environment. The problems created by industrialisation came to most people as a shock. Societies had optimistically and without any second thoughts assumed that nature could handle any activities the human mind could think of. There simply was no theory or attention to what negative impact the growth could have on nature. Up until now, nature was seen as an inexhaustible resource to be extracted and converted to the satisfaction of humans many needs and desires. What happened with waste from this process was not conceived as a problem. There was this innocence of unknowing the detrimental effects for a few decades.

In Denmark, the environmental mindset entered the public when dead lobsters showed up in the angler catches from the sea of Kattegat. They had suffocated due to lack of oxygen. Suddenly the public realised that the societal effect on the environment was not 'slightly detrimental', but rather catastrophic; something needed to be done. This led to the expansion of wastewater treatment plants handling both solids, organic matter and the key polluting nutrients nitrogen and phosphorous. In this move, utilities have changed from being merely an infrastructure provider to becoming a central environmental service in societies.

> **Your reflections:** If you were to understand this mindset better, who would you speak to? What would you ask?

## CHARACTERISTICS OF GRAVES MODEL

Graves theory is known as 'the emergent, cyclical, double-helix model of adult biopsychosocial systems development', better known as 'Spiral dynamics'. The progression through the mentioned mindsets can be organised according to a forward-moving spiral. The consecutive levels of patterns of thinking are organised on an expanding spiral so that, in a sense, for each turn of the spiral the previous levels are both encompassed and transcended. Each new level consists of a more encompassing and hence more complex way of thinking.

Three characteristic processes can be seen taking place in the spiral form of the process.

The first characteristic process is that the mindsets cycle between two extreme poles. One pole is a mindset that is moving against nature and the collective and its participants are very much focused on 'me and mine'. When this mindset is dominating (stages 1, 3 and 5), competition has priority over collaboration. At some point, however, this mindset leads to the beginning of disintegration of society, coherence suffers, societal values are under pressure. Then the spiral moves towards the opposite pole into harmony with nature and the individual's co-participation in its society; the focus changes to 'we and our'. These levels are characterised by 'higher ideals' and collaboration to build something larger together. At some point this becomes too restricted and oppressive for the individual to bear and the longing for individual freedom forces a change. In that sense, the spiral moves toward a stronger expression of self and its freedom for each level up along one side of the spiral. On the other side of the spiral, it moves into more and more harmonious integration of the collective. In a very simplified notion, it is a dialogue between safety together and freedom alone.

The second characteristic process is that the size of organisations the individual belongs to increases with each stage – as does complexity. At stages one and two, the attention revolves around the individual and a narrow family. At the third stage the group under control of a strong leader grows, however, it still depends on the strong individual's ability to manage the whole group by means of a few degrees of links from the strong leader. Everybody has a strong feeling of being part of an in-group against the rest of the world. At the fourth stage, the community makes a leap in size. Individuals now belong to large countries, ideologies and religions. This is a stage of great patriotism and loyalty to the cause and culture of the country or religion one identifies with. Still it is clear that there are the ones on our side and 'the infidels' on the other side, belonging to the other religion or the other country. In the sixth stage, the humanistic point of view tries to make away with all borders and boundaries toward equality for all, despite sex, race, religion etc. Thereby, the 'foreign' is embraced and attempted assigned equal rights, and there is this vision of one big (happy) humanity.

Hence, for each stage up, the individual's life-world span is enlarged. The individual's ability to mentally and emotionally increase his or her span becomes a precondition for the personal development along the spiral. Higher levels of mindsets on the 'we' part of the spiral have an expanding view of who 'we' are. Similarly, on the side of the 'me' side of the spiral where competition is more important, here the competition field expands from local towards global.

A third characteristic of the spiral is that the time at each stage becomes shorter and shorter. Beck and Cowan (2005) who interpreted Graves' model for practical use, tried to set a point in time where the evolutionary spearhead of humanity started thinking at a new stage. They came up with the following estimates. The beginning of the first stage lies 100–200,000 years back, the second stage lies

50,000 years back, the third stage has been around since possibly 10,000 years ago, the fourth stage rose 5000 years ago, the fifth stage started 1000 years ago and the latest stage to be reached by a significant amount of people, the sixth stage, was established 150 years ago. Today all the mindsets are present globally. The fifth stage mindset is dominant in leading countries like the US and Europe, followed closely in priority by the previous fourth mindset, and with the sixth 'green mindset' is on the rise. So there is a large time lag from when the first thinkers begin to explore new ways of thinking until it becomes part of mainstream society and even longer until society and individuals have arranged themselves so that acting from a more advanced stage becomes the dominant way of operation.

At some point the centre of gravity in a system (e.g. country, city, organisation, family, community) shifts. This is usually a noisy event regardless of whether the system is moving up or down the spiral. The Second World War could be used to illustrate this as an example of a red mindset (Hitler's Germany) battling a generally blue mindset (Churchill's United Kingdom). After years of fighting and an attack on Pearl Harbor, the orange mindset (US) came to the rescue from Roosevelt's United States bringing advanced technology and a scientific mindset. Reflections followed the warlike 'what could have moved the world into such a frenzy?', and not least 'what could lead such a large part of the German people into the arms of fascist thinking?' The results of these reflections led to a humanistically centred philosophy of unity and collaboration and the formation of 'green mindset' ideas such as the United Nations.

Every individual performs a personal development process going through corresponding personal levels of development. Starting from the beginning and landing at the level that matches the life conditions in which he/she finds themselves in terms of environment, family, network, culture, country, education, job etc. Hence, the development process can be understood as a way of understanding history, a way of understanding societies and a way of understanding the personal development processes and maturity levels of individuals.

Hence several mindset stages are present in each of us, and different mindsets may dominate in different areas of our lives. The mindsets are organised like the layers of an onion. The values of the individual generally gravitate toward the values of the group he or she primarily identifies with. As a rule of thumb, most people identify 50% with one level and 25% from a level above and 25% from a stage below the main step. Hence, generally, it is possible for most of us to understand mindsets that are further behind in the spiral and slightly ahead. It is however difficult or impossible from one's most advanced mindset to understand the mindsets ahead of this point on the spiral. Hence, people in the orange mindset will mostly scoff and shake their heads at the 'unrealistically idealistic naive types' of the green sixth mindset stage. People at the sixth mindset are known to scoff at every earlier mindset, seeing them as too simplistic. A key weakness in the green mindset is the strong assertion that everybody is equal, while at the same time they do not see people based in earlier mindsets as equal at all. People operating primarily from the red egocentric mindset tend to try to close all the long-winded discussions to '*just* go out and do something'.

When belonging to a 'higher' mindset level you might think that in 'the kingdom of the blind the one-eyed man is king', but that is not the case. Higher levels are not always advantageous. In general, the model suggests that the 'best' level of thinking is the one that replies best to the surrounding life conditions. Since being part of a 'human culture' constitutes a large part of a person's living conditions, most of the participants of a society will generally cluster around the same level since this for each individual is the most effective. The incentive to develop to reach that level in one's maturing process is high (or you will be excluded as an immature individual) however, the incentive to go higher than that shared level is low or negative. You may by such a step easily exclude yourself from your culture. To this Graves' comments:

> '*I am not saying in this conception of adult behaviour that one style of being, one form of human existence is inevitably and in all circumstances superior to or better than another form of human existence, another style of being.*
>
> *What I am saying is that when one form of being is more congruent with the realities of existence, then it is the better form of living for those realities.*'
>
> Clare W. Graves

When I look through the developmental levels of the model as described, I recognise periods in my life where each of them has been dominant and I am able to connect again to these earlier periods and thereby see the beauty and the courage of each level. For my life's journey, I think each level had its own importance, it was a prerequisite that I took on the set of values completely and wholeheartedly to exhaust the values completely in order to reach the limitations at each level.

A thing that often confused me is that at each stage of my life it was as if the words did not mean the same thing anymore. The model has helped me understand why this is so. The word 'courage' for example, as described at each level, has quite different meanings. From a red mindset of physical courage, you may not be able to recognise the courage at other levels as courage at all. You might even take it for cowardice. It is clear to me that when developing my patterns of thinking, each level has had its own type of courage to enable the knowledge or wisdom for the next stage to emerge.

At the third egocentric stage, the courage was about saying 'no' to being dominated, but instead to stand up against domination – and at times even being the dominating part. At the fourth absolutistic level, courage was, in a sense, to give up my self-interest for a higher purpose. At the fifth rational stage, it was about the courage and self-reliance to go outside the protection of convention and religion and to follow my own understanding, ambition and sense of what was right. At the sixth relativistic stage, it was the courage to step out of the competition and instead trust and lean into the community around me.

At each stage, my usage of each set of values felt 'through and through right – all the way into the marrow of me', making me feel on the 'right path'. The crises preceding each shift were slow and filled with moving forth and back in a regression-progression pattern and not an epiphany 'from one day to the other' kind of shift – though epiphanies did occur. But the process was more a number of slow realisations of something being off:

*'I do suggest, however, and this I deeply believe is so, that for the overall welfare of total man's existence in this world, over the long run of time, higher levels are better than lower levels and that the prime good of any society's governing figures should be to promote human movement up the levels of human existence.'*

Clare W. Graves

Seeing the world through the spiral perspective leads to a different understanding of many of the global conflicts and smaller conflicts in the local society, all the way down to conflicts in the workplace.

*'Different worldviews create different worlds, enact different worlds, they are not just the same world seen differently.'*

Ken Wilber (1996)

*Steven Solomon writes about the role of water in the process:*

*'Those unable to overcome the challenge of being farthest removed from access to the best water resources, by contrast, were invariably among history's poor.*

*History was littered with societies that declined simply because they could not overcome the deleterious local-resource depletions and population expansions accompanying their own initial success. Signature water challenges evolved from era to era.'*

Steven Solomon (2011)

The primary reason for water's importance in the developmental process is its fundamental nature. Without water, every stage that builds on top of this 'square one'-need crumbles. Steven Solomon orders the reasons for water's importance like this:

(1)  Domestically for drinking, cooking, and sanitation (stage 1 and 2);
(2)  Economic production for agriculture, industry, and mining (stage 5);
(3)  Power generation, such as through waterwheels, steam, hydroelectricity, and as a coolant in thermal power plants (stage 5);
(4)  For transportation and strategic advantage, militarily, commercially, and administratively (stage 3 and 4)
(5)  Of growing prominence today, environmentally to sustain vital ecosystems against natural and man-made depletions and degradations (stage 6).

Though he does not order the reasons according to spiral dynamics; the purposes can be recognised as belonging primarily to different stages of mindsets. Hence, water is clearly in dialogue with the evolution of new mindsets, sometimes enabling and at other times responding to needs.

The understanding of spiral dynamics represents a key pattern that enables a profoundly effective understanding of current conflicts and potentially also offers tools to resolve the conflicts. The ability to mediate these conflicts will be central for the global society's ability to move forward towards a sustainable future. These mindset conflicts are one of the major reasons for the lack of real progress in the area of sustainability. And many of these conflicts can be interpreted or traced back to the conflict between, especially, the four latest mindsets; the third stage egocentric, the fourth stage law and order, the fifth scientific and rational and the sixth green sustainability and equality focused stage. Generally, all levels in the first tier are in some kind of basic conflict. Only second tier mindsets are able to embrace and include all first-tier paradigms reliably and systematically. This emphasises the importance of SDG number 17 of global partnerships for sustainability. Hence, the understanding of these levels can prove central to securing water for all moving forward.

Even if the sixth green stage is not currently the dominant stage, it is already possible to see some of the shadow sides of this mindset, and it is evident that to truly transcend the current sustainability deadlock a mindset with more capabilities than the sixth stage is required. Some of the problems of the sixth stage are that it claims to be inclusive and regarding everyone as equal while holding a serious grudge against everybody, who doesn't see it like this. Plus, as anybody who has been involved in this type of 'green project', it is clear that 'green' has a hard time making decisions at all – which may create all kinds of problems on its own.

There are issues of great inefficiencies in decision-making processes as the mindset attempts to listen and include everybody convinced that it is possible to find a solution, that everybody finds acceptable and that harmony will emerge from that. When this idea meets the real world of a rainbow of mindsets, this is not a dead-sure result, obviously.

Beck and Cowan write in their book 'Spiral Dynamics' about the end of the sixth stage mindset: 'The beginning of the end of this value-memes dominance is when the person begins (again) to get things done, and done well, all alone. [...] A bit of disharmony becomes natural and one's tolerance for open contradictions grows. [...] Perhaps the most significant marker of the existing of GREEN/yellow [the exit of this stage] is the dropping away of fear. Life is life, after all. Tribal safety [stage 2], raw power [stage 3], salvation for all eternity [stage 4], individual success [stage 5] and the need to be accepted [stage 6] diminish in importance. Instead, there is a growing curiosity about just being alive in an expansive universe'.

## SECOND-TIER MINDSET

So what will be next?

Graves' model predicts that what happens next will be a momentous leap forward. Something fundamental happens in the shift from sixth to seventh stage. In Graves' model every mindset stage after the sixth stage are called second-tier mindsets – while stages one to six are referred to as first-tier mindsets. The major differentiator between first- and second-tier mindsets is that the second tier mindsets are not fear-driven anymore. Looking at Maslow's pyramid of needs, (Figure 16) there is a similarly remarkable shift when the individual has organised their ability to fulfil their needs on the lower levels sufficiently to such an extent that they start being occupied with self-realisation. This is predicted to be the most important enabler to move to the second-tier stages.

Key values and skills at the second-tier mindset are integrative, systemic, ecologic, authentic and better balanced. A distinctive feature is that at these levels, fear is almost absent. When looking at the world from a second-tier mindset, the world is conceptually different. Second-tier mindset has the key drive or purpose of stitching the world together, of overcoming the fragmentation and of enabling self-actualisation personally as well as collectively. This is a

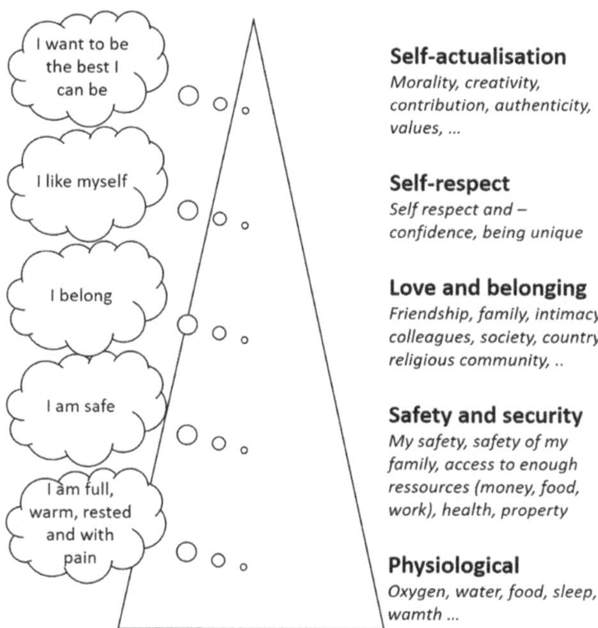


**Figure 16** Maslow's pyramid of needs. (*Source*: Adapted based on Maslow's pyramid of needs)

generous and open mindset where everybody is supposed to contribute based on their talents.

Earlier boundaries, delimiting one mindset from another or delimiting one country from another, in this new view, are transformed to something less fearful – something like mere landscape features.

> *'If the god is a tribal, racial, national or sectarian archetype, we are the warriors of his cause; but if he is a lord of the universe itself, men are brothers.'*
>
> Joseph Campbell (1949)

The realisation that becomes available to second-tier minds is that the joint and individual goals are not fundamentally in conflict.

An example of this change of mindset can be seen in the arena of environmental thinking. In this new mindset it becomes increasingly clear that taking steps to protect the environment is not – as earlier believed – a commercial hindrance. Instead it can finally be distinguished as a benefit for all. In this new mindset it is suddenly understood that environmental protection is both meaningful and valuable. What was earlier suggested from a primarily moral point of view, and hence consciously or subconsciously fought against because of the restrictions it proposes, can now be embraced because the 'meaningfulness' is clear. And the patterns can be enacted far more precisely when they are deeply understood rather than when they were interpreted from a dogmatic moral point of view, with its flavour of lifeless theory.

Hence, seen from the point of view of spiral dynamics, the question of this book may be re-formulated into: How will/can/should our relationship with water develop when we enter a second-tier mindset and how do we transcend our current mindset about our relationship to water into the framework of this new mindset – a new pattern of thinking about water? This kind of mindset will enable the solving of the current water crises.

> **Your reflections:** Does a second-tier mindset seem attractive? Which characteristics would you appreciate? There is some controversy as to whether one can move ahead alone, or it is more a thing that happen to you – what do you think? How would you describe your personal practice?

## CHANGE IN THE SCIENTIFIC PARADIGM

I was challenged by a teacher of mine to describe the difference between the old and the new mindset seen from a scientific point of view. At first the question puzzled me, but as I started making the list it soon became clear that though I had not thought of it that way, I knew of a distinct shift taking place between 'old-school' science and a 'new-school' science. By taking a step back and looking into the patterns of changes, some descriptive concepts can be identified as the old and the new paradigm. The list I generated based on my understanding and my reading was:

| The Currently Dominating Old Mindset | The Emerging New Mindset |
| --- | --- |
| The mechanical science paradigm | The systemic science paradigm |
| The study of objects | The study of life |
| Machine dominant as a metaphor | Organism dominant as a metaphor |
| Primary focus on quantity | Primary focus on quality |
| Linear | Non-linear |
| Simple – as billiard balls | Complex – as in biology |
| Isolation of phenomena | Phenomena are connected |
| Objects and affect | Structure and relationships |
| Utilitarian purpose | Purpose of understanding |
| The whole is a sum of the parts | The whole is more than the sum of the parts |
| Standardisation | Diversity |
| Material systems | Living systems |
| Shallow ecology | Deep ecology (Arne Naess) |
| Anthropomorphic (man in the centre) | Humans as one (highly developed species together with others |
| Nature has instrumental value | All life is valuable in itself |
| Hierarchy | Network |
| Reductionistic | Synthesis |
| Mental | Whole person view |
| The world as material | The world alive |
| Entropy | Homeostasis and autopoiesis |
| Precision | Patterns |
| One-to-one rules | Symbiotic fit |

The list was strongly inspired by Capra and Luisi's book the system's view of life (Capra & Luisi, 2014). Here they write:

> *'A central characteristic of the systems view of life is its nonlinearity: all living systems are complex – i.e. highly nonlinear – networks; and there are countless interconnections between the biological, cognitive, social, and ecological dimensions of life.'*
>
> Capra and Luisi

The new paradigm does not mark an abrupt departure from the old paradigm. In favour of the old paradigm, they write:

*'What makes the scientific enterprise feasible is the realization that, although science can never provide complete and definitive explanations, limited and approximate scientific knowledge is possible. This may sound frustrating, but for many scientists the fact that we can formulate approximate models and theories [..] is a source of confidence and strength.'*

Capra and Luisi

In regards to understanding what this means to our relationship with nature echoes the ideas of Naess as they assert:

*'Logic does not lead us from the fact that we are an integral part of the web of life to certain norms of how we should live. However, if we have the deep ecological experience of being part of the web of life, then we will (as opposed to should) be inclined to care for all of living nature.'*

Capra and Luisi

These change concepts can be interpreted in a number of different ways, examples of transformations taking place already that seem to work in the direction of this paradigm shift are:

### (1) An increasing role for biology

We have over the last decades seen an increase in the inclusion of biotech processes in technical water systems and our deepening understanding of the role of microbiology in our systems is a slow change of paradigm from a mechanistic perception to a more microbiological and life-based view on treatment.

For a long time bacteria were feared in drinking water, while now it is becoming more and more well known that bacteria are everywhere and that most are benevolent and we need only fear a subset of them.

Microbiology has been used in water treatment for a long time and is hence not necessarily a new thing, but what is new is the recognition of this and the study to understand the details of it. This is true in wastewater, but perhaps even more so in the increasing understanding of the microbiology involved in water treatment on the clean side. In the Tisso II water treatment plant it was clear that we had to understand microorganisms to be able to design a water treatment system without chlorine. The idea of chlorine in drinking water is in a sense basically an anti-life choice, where low-grade poison is added to the basic source of life to avoid bacteria.

What we understood during the Tisso II project was that bacteria played an important role in the water treatment process (sand filtration) and second, by creating a situation where the amount of nutrition for bacteria was removed it became possible to keep unwanted bacteria from the system without the use of chlorine.

However, this inclusion of biological solutions and learning from nature can be taken much further from here. New concepts such as biomimetic membranes using aquaporin molecules similar to those in cell membranes and the development of sensors for detecting harmful bacteria by means of DNA sequencing methods signals a new world. In this new world operation can become more precise and flexible and hence avoid making the system 'over-safe' as in the case with chlorine. Other emerging scientific options are quorum sensing. The way that microorganisms communicate is believed to increase our knowledge of how to improve the beneficial circumstances for microorganisms, the use of enzymes to tackle difficult substances in water provides new possibilities, the use of microalgae for wastewater treatment makes the process regenerative rather than consuming.

The inspiration from the biomimicry movement is promising. According to the inventor of the term 'Biomimicry', Janine M. Benyus (2002), the nine basic principles of biomimicry are:

(1)   Nature runs on sunlight
(2)   Nature uses only the energy it needs
(3)   Nature fits form to function
(4)   Nature recycles everything
(5)   Nature rewards cooperation
(6)   Nature banks on diversity
(7)   Nature demands local expertise
(8)   Nature curbs excesses from within
(9)   Nature taps the power of limits

Wetlands is an example of an ecosystem that has been used as inspiration for a number of different inventions. Based on the study of wetlands biomimicry solutions for ecosystem solutions, filtering technologies, desalination, flow technologies and surface protection were identified (Water Research Commission, 2013).

## (2) Understanding of eco-system requirements

The understanding of eco-systems requirements for survival thriving is deepening. During the 1980s and 1990s, the understanding of the nutrients nitrogen and phosphorous became well understood and legislation was implemented accordingly. Making the wastewater systems live up to this new knowledge took several decades. But still many places in the world have not reached that level yet. In the last decade, the understanding of other substances in water has made the alarm bells ring: pesticides, persistent organic pollutants (POPs), herbicides, pharmaceuticals, antibiotics, high production volume chemicals, heavy metals, emerging pathogens, endocrine substances, microplastics etc. These are all substances that have been used in society for decades but are now accumulating to a point where their effect on the health of the aquatic ecosystems is threatening and hence also becomes threatening for our human health situation.

Ecology and the deepening of the ecological understanding is one key to the recent incline in popularity of the concept of sustainability. The new insights into the life of trees and their underground collaborative communication network of mycelium spurs a new research interest and appreciation of the intricacy of our biodiverse surroundings.

An interesting recent development is the granting of legal personhood to rivers. Now the rivers Whanganui River in New Zealand, Ganges and Yamuna Rivers in India, Vilcacamba River in Ecuador and Atrato River in Columbia have been granted some legal protection as if they were humans. This is a promising way of ensuring 'water stewardship'. By this method the rights of rivers are codified, putting an end to the rights to over-exploit them by private companies or through public permits. Similar 'personhood' rights are considered for trees, forests and other natural entities. This may be an improved way of protecting natural reserves. An alternative future approach may entail an 'ombudsman for water'.

### (3) The smart revolution of digitalisation

The revolution of, for example, Smart Water Utilities, has taken place since the invention and proliferation of computers since the 1970s and 1980s. The development speed in those decades is amazing. In a sense, it is building a new layer of intelligence on top of the existing steel and concrete systems for water and wastewater treatment, distribution and collection. The development of this new level of intelligence is rooted in a number of inventions:

- Instrumentation, sensors and actuators
- Sensor validation
- Monitoring and data validation
- Telemetry and communications
- Data and information management
- Process control
- Standardisation of protocols
- Modelling of the network, treatment systems
- Etc.

Today the concepts of digital twins, Big data and industry 4.0, are key issues:

Digital twin: the idea that the urban water cycle is mirrored digitally. This enables one-to-one modelling and scenario testing, systematic and comprehensive maintenance management systems and running design optimisations of the fully integrated systems.

Big data: The application of the enormous amount of data accrued during operation. This is especially about mathematical algorithms for the identification of high-level patterns in large amounts of data; patterns beyond what can be perceived and identified by human operations.

Industry 4.0: the latest paradigm shift in how industrial systems are designed. Industry 1.0 refers to the beginning of the industrial revolution focused on the use of steam and water power. Industry 2.0 refers to the period where electricity was key to cutting edge technology. Industry 3.0 was marked by the introduction of computers and especially digitalisation. Industry 4.0 is primarily related to the invention of the internet and mobile devices, internet of things, smart sensors, cloud computing and data visualisation. It is guided by principles of interconnection, transparency and decentralisation – all aspects that are relevant in the 'new paradigm' or the new story.

This industry 4.0 concept has not yet found its final form, and the water and wastewater utility industry are not necessarily leading this development. But it affects the industry enormously and has become an aspirational term that water engineers are struggling to give a real effective beneficial form.

When looking at the above examples of changes from the old paradigm to a new one it is clear that neither of the three examples represents a full transformation. They are all tentative attempts and have a strong pioneering quality to them.

Second, there are a number of other areas within water that need to be addressed, and where the change is hardly discernible, for example:

### (4) Women in water

Women in water have been a theme for a while. It is a theme in water journals and it has been described as an important aspect of IWRM. Women have a different approach to water, but are still not representing a strong voice. This is not just a matter of the number of women in water engineering. It also has to do with including a more feminine take on water – which both men and woman can. Hence it is also a challenge for women to truly find their voice about water rather than continuing the current 'masculine' approach. An inspiring example is listening to the way Indian Philosopher Vandana Shiva argues for a more ecological approach, not just to water but to life on all levels. An example of this is her nine rules for water democracy, where she couples the usage of water with natural principles and principles of fairness and democracy (Shiva, 2007):

(1) Water is a gift of nature: We receive water gratuitously from nature. It is our duty towards nature to use this gift according to our needs and sustenance, keep it clean and in adequate quantity. The deviations which create arid or flooded regions violate the principle of ecological democracy.

(2) Water is essential to life, water is the source of life for all species: all the species and all the eco-systems have the right to their share of water on the planet.

(3) Life is interconnected by means of water: water connects all human beings and every part of the planet through its cycle. We all have the duty to ensure that our actions do not provoke damages to other species and to other persons.

(4) Water must be gratuitous for the needs of sustenance. Since nature grants us the gratuitous use of water, to buy or to sell it in order to make a profit, violates our inner right to the gift of nature and detracts from the poor their human rights.

(5) Water is limited and subject to exhaustion. Water is limited and can be exhausted when used in a non-sustainable manner. To draw more water from the eco-cistern

than nature can furnish is an unsustainable way of using water and consuming more than one's own legitimate share, since other's have the right to a fair share.

(6) Water must be preserved: Everyone has the duty to preserve water and to use it in a sustainable manner with the sustainable and just limits.

(7) Water is a common good: Water is not a human invention. It cannot be confined and it has no confines. It is by nature a common good. It cannot be possessed as private property and sold as merchandise.

(8) No one has the right to destroy it: No one has the right to make excessive use, abuse, waste or pollute the system of circulation of water. The permits of commercially allowed pollution violate the principle of just and sustainable use of water.

(9) Water is not substitutable: Water is intrinsically different from other resources and products. It cannot be treated as merchandise.

Vandana Shiva

### (5) Flexibility in infrastructure design

Most water infrastructure designs are made too large and inflexible – often on multiple levels: for example, capacity of plants often is much larger than needed, individual components such as pumps and compressors which tend to operate far away from optimum and are difficult to regulate or the relevant range and the variable range doesn't fit, because they are chosen for the extreme case. When the system is designed with some flexibility, the provided flexibility is often not used correctly because of lack of control competence.

A silently emerging trend is the use of smaller water treatment units, that fit into truck containers. Containerised solutions will, over time, reach a maturity level that provides an option of more flexible increasing and decreasing capacity. The containers enable water providers to expand and decrease capacity as demand changes. Also smaller household units are emerging for houses and apartment buildings. Still, when looking at the multiple solutions found in nature, I imagine even more flexible solutions can be developed – a recent example is machines for harvesting atmospheric vapour from the air.

These are just some of many areas of our engagement with water needs to or about to change. These are truly interesting times – but it also requires something of all of us to make these changes happen.

---

**Your reflections:** Do you see any of these 'trends' play out in your work? How do you practically engage with these new trends? What emerging trends can you identify besides the above that are important in your work?

# *Chapter 4*

# Potential frameworks

---

The term 'water stewardship' was first used in writing in 1970 and has since increased in usage. It is, however, still a rare word, as can be seen in Figure 17.

The graph in Figure 17 shows how often the word occurs in the English literature. The ordinate axis shows the percentage of the occurrence of the word as to every other word – which as can be seen is staggering low. To understand how rare the word is I attempted to find words that had the same rarity. I found the following words of similar narrow use: ailurophile (cat lover), cereology (the study of corn circles) and discobolus (a discus thrower in ancient Greece). I didn't know any of the words and had to make a Google search for 'rare words'.

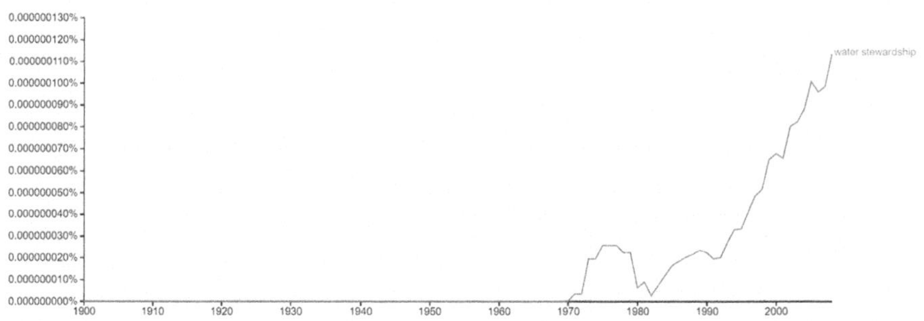


**Figure 17** The development of the use of the word 'water stewardship' as per the Google Ngram Viewer software. (Michel *et al.*, 2011)

© IWA Publishing 2020. Water Stewardship
Author: Pernille Ingildsen
doi: 10.2166/9781789060331_0121

So in general, while the word 'water stewardship' is increasingly popular, it is still a very rare word.

The etymology service Etymonline.com writes about the origin of the word stewardship as follows:

*Steward: Old English stiward, stigweard 'house guardian, housekeeper,' from stig 'hall, pen for cattle, part of a house' (see sty (n.1)) + weard 'guard' (from Proto-Germanic *wardaz 'guard,' from PIE root *wer- (3) 'perceive, watch out for').*

*Used after the Conquest as the equivalent of Old French seneschal (q.v.). Meaning 'overseer of workmen' is attested from c. 1300. The sense of 'officer on a ship in charge of provisions and meals' is first recorded mid-15c.; extended to trains 1906. This was the title of a class of high officers of the state in early England and Scotland, hence meaning 'one who manages affairs of an estate on behalf of his employer' (late 14c.). Meaning 'person who supervises arrangements' at a meeting, dinner, etc., is from 1703.*

*Stewardship: 'position or responsibilities of a steward,' mid-15c., from steward (n.) + -ship. Specific ecclesiastical sense of 'responsible use of resources in the service of God' is from 1899.*

The Merriam-Webster dictionary explains stewardship as *the conducting, supervising, or managing of something, especially: the careful and responsible management of something entrusted to one's care.*

Dictionary.com defines stewardship as *the responsible overseeing and protection of something considered worth caring for and preserving.*

*RELATED WORDS: control, management, supervision, protection, conservancy, maintenance, care, sewing, housekeeping, keeping, safekeeping, saving, attention, storage, governing, upkeep, economy, salvation, custody, guardianship.*

The Alliance for Water Stewardship, the World Wildlife Foundation and the Water Footprint Network are all organisations that have adopted the concept of water stewardship and hence made an effort to attempt a definition.

> *'Stewardship is about taking care of something that we do not own. Good water stewards recognise the need for collective responses to the complex challenges facing the water resources we all rely on. Good water stewards understand their own water use, catchment context and shared risk in terms of water governance, water balance, water quality and important water-related areas; and then engage in meaningful individual and collective actions that benefit people and nature.'*
>
> Alliance for Water Stewardship

> *'We define Water Stewardship for business as a progression of increased improvement of water use and a reduction in the water-related impacts of internal and value chain operations. More importantly, it is a commitment to the sustainable management of shared water resources in the public interest through collective action with other businesses, governments, NGOs and communities.'*
>
> World Wildlife Foundation

*'Concern about water in the private sector is rising rapidly. Companies worldwide are recognising that water is an essential ingredient in their business operations and the lack of access to sufficient water quantities or quality is posing a material risk to a growing number of companies. Water stewardship is one way a company can manage the complexities of balancing their own water use with the needs of communities and nature.'*

Water Footprint Network

So while the definitions vary, there is the same common intuitive feel of the meaning of the word, including:

- Taking care of something that is a collective resource and not owned by you;
- In service to something larger than oneself and human society;
- Sustainable use, protection and overseeing of water;
- Integrating and mediating the complexity of different stakeholders needs for water;
- A special responsibility to nature's requirement for water.

Water stewardship can be used as an aspirational term. The purpose of this chapter is to try to capture the spirit, aspiration and intention of water stewardship in order to move us into a 21st-century role. It is a movement from water professional to water steward and a movement from managing water to co-using water.

A water stewardship approach would be an upgrade of water institutions such as utilities, municipalities, technology vendors, water consumers and research institutions to take part in a joint responsibility for water stewardship.

## INTEGRATED WATER RESOURCE MANAGEMENT

Integrated Water Resource Management (IWRM) is a term that has a meaning aligned with water stewardship. The Global Water Partnership uses the definition 'IWRM is a process, which promotes the coordinated development and management of water, land and related resources in order to maximize the resultant economic and social welfare in an equitable manner without compromising the sustainability of vital ecosystem' (Hassing *et al.*, 2009). IWRM is focused on balancing economic, social and environmental aspects around water and land. IWRM have a freshwater resource focus and seem primarily occupied with surface water resources (though groundwater is also mentioned).

The concept of IWRM was developed over a number of key conferences (Rahaman *et al.*, 2004):

International Conference on Water and Environment in Dublin in 1992: at this conference, it was recognised that IWRM takes place at three scales: local, national and international. The conference suggested the following four principles (WMO, 1992):

(1)  Freshwater is a finite, vulnerable and essential resource, which should be managed in an integrated manner.

(2)  Water development and management should be based on a participatory approach, involving users, planners and policymakers at all levels.

(3)  Women play a central role in the provision, management and safeguarding of water.

(4)  Water has an economic value and should be recognized as an economic good, taking into account affordability and equity criteria.

Second World Water Forum and Ministerial Conference in The Hague 2000. This conference dealt with privatisation and warned about substituting a government monopoly with a private monopoly. The forum discussed the cost of water. The idea was that charging the full cost for water services would increase the incentive for water savings and this would help the financial requirements to ensure universal access to water. Additionally, appropriate subsidies should be made available to the poor. The rights to use of water and land for agriculture for the poor was also discussed (World Water Council, 2000)

International Conference on Freshwater in Bonn, 2001: At this point, it was clear that IWRM was not as easy to implement as expected. Hence, the conference focused on guidelines of practice. It was realised that the first issue to solve was to meet water security needs for the poor; hence, decentralisation and partnerships came into focus. It was realised that the central unit in IWRM processes are water basins, and that a strong governance framework would be important to ensure that the collaborative process would succeed (IISD, 2001).

The World Summit on Sustainable Development in Johannesburg in 2002: Here, the principles earlier laid out earlier were elaborated upon and with an increasing focus on sustainability. There was a pressure to develop IWRM plans for major water basins on a short time horizon making use of public–private partnerships and involving all concerned stakeholders. Additionally, there was a special focus on women and gender-sensitive programmes (IISD, 2002).

Throughout these conferences there seems to be the main focus on developing countries, though this is not explicitly stated. Regardless – and this is important – it becomes increasingly clear that living up to the IWRM principles of gender sensitivity, decentralization, participation of stakeholders, focus on poverty, how to include the private sector in the discussions, ensuring coordination across stakeholders, focusing both on technology and human management, is much more difficult than originally expected. When experts and policymakers meet, there is clearly a shared logic as to how water resources should be managed. In practice, however, it is difficult.

The Global Water Partnership (GWP) is an international network whose vision is a water-secure world. The GWP mission is to support sustainable development and management of water resources at all levels. It was created in 1996 to promote IWRM practices and in 2009 they published a handbook in IWRM. The handbook addresses the large-scale challenge of IWRM and states that: 'On the one hand, water is essential to human, animal and plant life. Water supports productive activities, agriculture, generation of hydropower, industries, fishing, tourism, transport, for example. On the other hand, water can be extremely destructive, carrying diseases and flooding vast areas. Insufficient water or prolonged drought can result in widespread death and economic decline. Water can also cause or escalate conflicts between communities in a local or national basin, or in transboundary basins shared by more than one country' (GWP, INBO, 2009).

'The concept of IWRM though in principle right poses a number of challenges as river basins, does not follow country boundaries and governance structures are not set up to handle these challenges. The handbook is structured around some of the key challenges:

- What political and legal factors do basin managers need to understand and take into account?
- What are the functions and what are the different kinds of institutional and legal arrangements for basin organisations?
- What are the different ways in which basin organisations and basin management can be financed?
- What type, level, structure and frequency of stakeholder involvement should basin managers seek to establish?
- How should basin managers go about strategic planning?
- What do basin managers need to consider in developing and implementing basin action plans, and how can they get feedback on how plans are progressing?
- What data and information management systems do basin managers need for integrated water resources management?
- What are the key communication issues basin managers need to consider?'

Global Water Partnership (GWP, 2009)

One finding, that I strongly agree with, and that I find important to underline, is that IWRM is a continuous and iterative process that is centred around a process of 'learning by doing' – which is similar to the natural process of trial and error. The ideal of local stakeholder-based management is easily overridden as the political and technical complexity is high – especially in these basin-wide systems spanning hundreds of square kilometres.

In a GWP background paper (GWP, 2000) on IWRM, the most important challenges that IWRM tries to overcome are stated as:

- Securing water for people (still a large part of the global population is without access to clean water);
- Securing water for food production;
- Developing other job-creating activities;
- Protecting vital eco-systems;
- Dealing with the variability of water in time and space;
- Managing risks;
- Creating popular awareness and understanding;

- Forging the political will to act;
- Ensuring collaboration across sectors and boundaries.

Many of these challenges can be recognised as topics from the comparable micro-level project around Tisso. It is clear that a holistic approach is required, it is also clear that there are nature-given limits to the resources and that the resources cannot be taken up entirely by human needs, but needs to be shared to sustain natural habitats and that the actual governance structure for facilitating IWRM is non-trivial. The IWRM has suffered from, in spite of intentions, being more top-down than bottom-up. Additionally, the focus has been primarily on the water supply side of the water issue, while the handling of wastewater has received considerably less attention. The focus of IWRM has been primarily on distribution and not so much on the actual state of basins and aquifers.

> **Your reflections:** Where in your work can you apply some or all the principles of IWRM? Where have you applied them already?

## THE ALLIANCE FOR WATER STEWARDSHIP

The approach by the Alliance for Water Stewardship may offer a complementary approach to IWRM. The method in the approach to water stewardship proposed by the Alliance for Water Stewardship is closer to a bottom-up approach. The Alliance for Water Stewardship is a global membership organisation including businesses, NGOs and the public sector. The organisation attempts to start a movement where members, through the adoption of the certification system 'the International Water Stewardship Standard', increase their awareness of own water issues as well as work on increasing their understanding of the greater water system they are part of. The approach embraces the 'learning-by-doing' principle by including a systematic approach to continuous actions working with water in an integrative way.

A fundamental principle is the belief that if major water users gain a sufficient understanding of their own water use and impacts on water in nature, and commit to publish these results transparently, then they will find the will and ways to improve. The public element and some of the criteria for the certificates ensure facilitation of coordination with other water users within the wider water catchment area.

The water stewardship standard (AWS, 2019a, b) covers the areas of good water governance, sustainable water balance, good water quality, important water-related areas and safe water, sanitation and hygiene for all (WASH), see Figure 18.

To have 'good water governance' means to have a governance system implemented so that responsibility is clarified and a framework for action is

**Figure 18** Key focus areas in the water stewardship standard. (AWS, 2019a, b)

implemented internally. This framework for actions must include a system for maintaining and updating water sustainability-related information.

'Sustainable water balance' means that the user's direct water consumption is withdrawn from natural resources in a way to ensure that the water resources are continually replenished. Similarly, the consumer needs to look into the sustainability of its significant indirect water use when possible.

'Good water quality status' means understanding and taking co-responsibility for the water quality of the aquifer from where the water is abstracted.

The topic of 'Important water-related areas' is about identifying impacted areas with special water interests, either for water abstraction or wastewater effluents. In these areas, the water steward must comply with effluent standards as well as engage in the quality of the water-related area as a prerequisite for thriving eco-systems.

'WASH' is shorthand for safe water, sanitation and hygiene for all and represents the social responsibility for other less powerful water users. The criteria are related to the concept of 'water as a basic human right'.

The process of water stewardship is an iterative cyclical process in five steps, as shown in Figure 19:

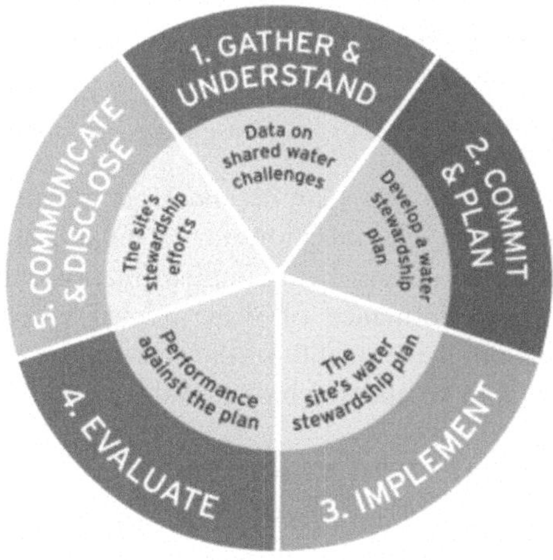



**Figure 19** The process of water stewardship. (AWS, 2019a, b)

## Step 1 Gather and understand

In this step, the operator conducts a comprehensive data collection to obtain an overview of the actual water situation. The standard specifies the type of data that has to be gathered and analysed, including:

- Outlining the geographical scope of the entities own operation;
- Identify stakeholders and their water-related challenges and mutual impacts;
- Collect key water information on water balances, water quality, cost and governance;
- Understanding own indirect water use;
- Understanding current and future water challenges in the geographical area;
- Understanding water risk and opportunities;
- Overview of best water-related practice.

## Step 2 Commit and plan

This step is about making the commitment to water stewardship public and practical. The step includes:

- The senior-most manager in charge of water must sign and publicly disclose the commitment;
- Develop an internal process for water stewardship in the organisation;
- Create a water stewardship strategy;
- Demonstrate the site's responsiveness and resilience to water risks.

## Step 3: Implement

In this step, the actions in the action plan are implemented including best practices for water stewardship. The actions relate to areas of water balance, water quality, important water areas, WASH, shared resources, as well as documenting legal compliance to relevant legislation water authorities.

## Step 4: Evaluate

The evaluation step is similar to the 'plan-do-check-act' process cycle seen in most ISO standards (International Organisation for Standardisation) and ensures a continuous evaluation of the effects of the implementation stage in step 3. It also includes evaluating the performance of water-related emergencies if any has occurred and an evaluation of the quality of the stakeholder dialogue. Finally, the water stewardship plan is to be updated based on the evaluation.

## Step 5: Communicate and disclose

Here the water stewardship information such as governance system, plan, performance against targets, effort and results of stakeholder collaboration and compliance etc. are made public.

For each step of the process, the organisation must comply with a number of criteria; to give an impression of these criteria, some examples are (AWS, 2019a, b):

- Criteria 1.1: Gather information to define the site's physical scope for water stewardship purposes, including: its operational boundaries; the water sources from which the site draws; the locations to which the site returns its discharges; and the catchment(s) that the site affect(s) and upon which it is reliant (this is an example from step 1: gather and understand).
- Criteria 2.3: Create a water stewardship strategy and plan, including addressing risks (to and from the site), shared catchment water challenges, and opportunities (this is an example from step 2: commit and plan).
- Criteria 3.1: Implement a plan to participate positively in catchment governance (this is an example from step 3: Implement).
- Criteria 4.1: Evaluate the site's performance in light of its actions and targets from its water stewardship plan and demonstrate its contribution to achieving water stewardship outcomes (this is an example from step 4: Evaluate).
- Criteria 5.3: Disclose annual site water stewardship summary, including the relevant information about the site's annual water stewardship performance and results against the site's targets (this is an example from step 5: communicate and disclose).

The certification comes in three grades: core, gold and platinum, depending on the number of indicators the organisation meets.

I appreciate how the system encourages collaboration with other partners through collective action with other stakeholders in the same water area, as well as organisations representing indirect water use. At the same time, this may be a way of extending the network of organisations to obtain compliance to the water stewardship standard.

The framework of the Water Stewardship standard is comprehensive and integrates the actions required to address the complex issues of water to ensure good water governance. The standard seems first and foremost directed towards large industrial water consumers but ought also to give, for example, utilities or municipalities a firm grasp on all their water issues. According to the Alliance for Water Stewardship, the organisations that are currently certified according to the water stewardship standard primarily include the business, agriculture and health care sectors and a few cities.

The framework of water stewardship certification has several similarities with more well-known standards like the ISO 9000 (the International Organization for Standards standard on Quality Management) and even more relevant the ISO 14001 (Environmental Management System, EMS). For utilities, the Water Stewardship standard can in principle be used as an effective framework for keeping track of water sustainability in all its aspects as part of the ISO 14001 or ISO 9000 standards.

While this approach in principle solves some of the difficulties with IWRM, such as a stakeholder-based approach that is more bottom-up than top-down, it does also hold some challenges. First of all, it (at least so far) addresses primarily major water users and not the multitude of other water stakeholders in an area. Second, it may be very difficult from a water user perspective to find solutions in a water basin because the consumer does not hold any formal power over what everybody else does. Third, the certification process is voluntary and will only be carried out to the extent that a given company takes an interest in doing this kind of work.

> **Your reflections:** What do you see as the key advantages and disadvantages with a standard certification approach? How much water sustainability mapping have you already carried out? Do you and your colleagues have a shared understanding of the water challenges around you, where you have a direct or indirect influence?

## WATER FOOTPRINT

The Water Footprint Network promotes the concept of water footprint. A water footprint can be calculated for a product, an organisation or a country. Water footprints are divided into three different types of footprints (WFN, 2019):

(1)  The green water footprint is water taken up from the root zone of soil, typically by agricultural products.
(2)  The blue water footprint is water abstracted from groundwater or surface water sources and used in various typically industrial processes. This is, for example, the water abstracted by utilities.
(3)  The grey water footprint is the amount of freshwater required to assimilate pollutants to meet specific water quality standards.

The water footprint method includes direct and indirect water consumption. The network has carried out some elaborate work analysing a number of different water footprints:

- The production of one kilogramme of beef requires approximately 15 thousand litres of water (93% green, 4% blue, 3% grey water footprint). There is a huge variation around this global average. The precise footprint of a piece of beef depends on factors such as the type of production system and the composition and origin of the feed of the cow.
- The water footprint of a 150-gramme soy burger produced in the Netherlands is about 160 litres. A beef burger from the same country costs on average about 1000 litres.

- The water footprint of Chinese consumption is about 1070 cubic metres per year per capita. About 10% of the Chinese water footprint falls outside China.
- Japan with a footprint of 1380 cubic metres per year per capita, has about 77% of its total water footprint outside the borders of the country.
- The water footprint of US citizens is 2840 cubic meter per year per capita. About 20% of this water footprint is external. The largest external water footprint of US consumption lies in the Yangtze River Basin, China.
- The global water footprint of humanity in the period 1996–2005 was 9087 billions of cubic meters per year (74% green, 11% blue, 15% grey). Agricultural production contributes 92% to this total footprint.

Water Footprint Network (WFN, 2019)

The Water Footprint Network provides tools to calculate the water footprint of individuals (simplified and extended version), a water footprint assessment tool and a national water footprint explorer providing data on the water footprint of most countries, including information on the distribution between internal and external consumption. Finally, they provide a list of products for which the water footprint has been analysed.

Additionally, for more professional use, the Water Footprint Network provides tools to analyse the water footprint of companies.

> **Your reflections:** How much do you know about your water footprint? Do you have a sense of the water footprint of various products or is it completely unknown to you? Were you aware of the three different kinds of water footprints?

## WATER STEWARDSHIP IN THE SUSTAINABLE DEVELOPMENT GOALS

The sustainable development goals (the SDGs) (UN, 2015a, b) represent a different framework for implementing water stewardship. For most water-related organisations there are many options for how to implement the SDG framework and typically more of the goals will be relevant.

Before applying the SDGs it is helpful to understand the full context of the SDGs. Otherwise, one may easily find oneself in a strange terrain, where the SDG goals and indicators are taken too literally. Organisations trying to translate the global context in a one-to-one way to the local level, attempting to influence the indicators directly in a strangely competitive way, will not achieve the success they hope for. Instead, any discussion about SDGs in organisations should be started by reminding everybody about the two basic concepts in the SDGs: 'leave no one behind' and 'do no harm'.

The SDG is an impressive human innovation born in the UN in 2015 with goals reaching the year 2030. The previous UN goal framework, the Millennium Goals (WHO, 2000), were primarily aimed at the then-called developing countries. In the 1970s the world was divided into developed and developing countries. This division is not so relevant anymore as all countries today are distributed more evenly on a continuum from most to least developed, and many previously developing countries have effectively transcended that status.

The SDGs represents a significant change of mindset from the Millennium Goals. The SDGs are goals for the whole world. With this change of mindset, every country in the world could be said to have become a 'developing country'. The SDGs succeeds in conceptualising that all the goals are interconnected and that we live in an interconnected world. Improvements in one area may lead to improvements or deteriorations in another area. The goal is to find solutions for one that does not sacrifice the other and preferably supports other areas. Aim always for solving more than one problem.

This change in mindset has come as a continuation of the increasing understanding of the world as one finite interconnected place. An idea that was rooted in the 1970s by the Club of Rome and moved on by researchers and philosophers such as Rachel Carson, Buckminster Fuller, James Lovelock, Lynn Margulis, James Hansen, Vandana Shiva, Charles Eisenstein, Wendell Berry, Al Gore, E. F. Schumacher and many more.

The central goal in regard to water is of course SDG goal 6, 'Ensure availability and sustainable management of water and sanitation for all'. Each goal consists of the number of targets and indicators, for SDG 6 they are:

**Goals**

- 6.1 By 2030, achieve universal and equitable access to safe and affordable drinking water for all.
- 6.2 By 2030, achieve access to adequate and equitable sanitation and hygiene for all, and end open defecation, paying special attention to the needs of women and girls and those in vulnerable situations.
- 6.3 By 2030, improve water quality by reducing pollution, eliminating dumping and minimizing release of hazardous chemicals and materials, halving the proportion of untreated wastewater and substantially increasing recycling and safe reuse globally.
- 6.4 By 2030, substantially increase water-use efficiency across all sectors and ensure sustainable withdrawals and supply of freshwater to address water scarcity, and substantially reduce the number of people suffering from water scarcity.
- 6.5 By 2030, implement integrated water resources management at all levels, including through transboundary cooperation as appropriate.
- 6.6 By 2020, protect and restore water-related ecosystems, including mountains, forests, wetlands, rivers, aquifers and lakes.

- 6.a By 2030, expand international cooperation and capacity-building support to developing countries in water and sanitation-related activities and programs, including water harvesting, desalination, water efficiency, wastewater treatment, recycling and reuse technologies.
- 6.b Support and strengthen the participation of local communities in improving water and sanitation management.

(UN, 2015a, b)

> **Your reflections:** How are the SDGs relevant to what you do? Are you happy with SDG6 or do you think something is missing – can you find it in any of the other SDGs? Do you know how well SDG6 is progressing?

## WATER AS A COMMON GOOD

The word 'common good' is related to the concept of commons, most notably known from the 'tragedy of the commons'. The 'tragedy of the commons' was described in 1968 by Garret Hardin in an article by the same name (Hardin, 1968).

Hardin explains the 'tragedy of the commons' with grazing a pasture. Having more herdsmen using the open pasture to keep cattle may work reasonably well for centuries as tribalism, poaching and disease keep the number of herdsmen and beasts below the critical level of the carrying capacity of the pasture. However, as the struggling society finally finds a way to reach a level of civilisation where peace and prosperity rules, the problem of the common good becomes apparent. The logic of sharing a resource, the common good, develops into tragedy.

The reason is that each herdsman will try to optimise his gain. When considering adding one animal to his herd, his calculations look like this. There is both a gain and a loss. The gains from the added animal are his alone. The loss due to overgrazing, on the other hand, is shared between all the herdsmen. Hence, he will have a net benefit by adding an animal to his herd. As time passes and every herdsman have this same logic, overgrazing becomes an increasing problem until the ecological limitations of the pasture breaks-down and everybody, herdsmen and animals suffer. And that is precisely the 'tragedy of the commons':

> *'Each man is locked into a system that compels him to increase his herd without limit – in a world that is limited. Ruin is the destination toward which all men rush, each pursuing his own best interest in a society that believed in the freedom of the commons.'*

Hardin (1968)

The 'tragedy of the commons' is acutely relevant when it comes to water, both in terms of allocation of the resource of freshwater and water's capacity for handling wastewater residuals. Therefore a key responsibility of the ideal water steward is to ensure allocation of this common resource. This should be done under the headline of *'the careful and responsible management of something entrusted to one's care'*. The 'something entrusted' is water, but by whom has the water steward been entrusted the careful and responsible management of water?

Water as a common good is relevant to all living, and it must, therefore, be 'entrusted' on behalf of 'all living'. Hence the responsibility expands beyond human needs to the lives of all creatures. Water as an absolute essential prerequisite for life is a commons belonging to all living. How can this be managed responsibly?

Hardin begins his article with a comparison with the problem of an arms race between nations. He pays tribute to the bold results presented in a paper by Weisner and York (1964). After they had looked carefully into the arms race dilemma, they concluded that there is no technical solution to the problem. As one country increases the number of arms, the other country will feel a decrease in national security, and in response will increase its own number of arms. This leads to an ever-escalating arms race and Weisner and York warned about impending nuclear war. However, what Hardin finds interesting is that there is no 'technical solution' to the problem. It is like playing tick-tack-toe. If both players understand the game and play within the conventions, none of them will ever win.

Hardin's conclusion on the 'tragedy of the commons' was that the application of organising rules imposed from above was necessary to handle these types of dilemmas.

Economist Elinor Ostrom, winner of The Sveriges Riksbank Prize in Economic Sciences in Memory of Alfred Nobel 2009 (Nobel prize in Economics), upon a lifetime of studies, came to a different conclusion.

She received the prize for the contribution of 'Challenge the conventional wisdom by demonstrating how local property can be successfully managed by local commons without any regulation by central authorities or privatization'.

Instead of a call for top-level management, she advocates that polycentric systems (that is systems with multiple centres) are more effective in handling commons than top-level centralised units (Ostrom, 2009). In an in-depth study of a police force divided into multiple different-sized departments in metropolitan areas, she found that there was not a single incidence where a large centralized police department outperformed smaller departments serving smaller neighbourhoods; in regard to multiple indicators. Similarly, from a study of irrigation systems in Nepal that compared 'engineered and run by government'-systems with 'built and run by farmers'-systems, she found that the latter type of system was usually more primitive, but they were able to grow more crops, run the systems more effectively and get more water to the tail-end of the irrigation system.

A forest study of hers found that users monitoring forests were more important to the thriving of the forest than the type of forest ownership. The care was even stronger when local communities had strong rule-making autonomy and incentives to monitor.

She found that effective systems usually included:

- Communication among participants;
- The reputation of participants is well-known;
- Longer time horizons;
- Agreed upon sanctioning mechanisms;
- Conflict resolution mechanisms in place.

The most important value in these polycentric networks, she found, was trust. Elinor Ostrom posits that 'all factors that increase the likelihood that participants gain trust in others and reduce the probability of being a sucker' increase the effectiveness of poly-centred systems. A central take-home message from her research is that we need to learn to deal with complexity rather than rejecting it. And exactly handling complexity is what polycentric systems are good at (Ostrom, 2009)

> **Your reflections:** Do you agree with Hardin or Ostrom? What do the two viewpoints mean practically in your work? What future would you like to live in?

# IN SUMMARY

While the concept of water stewardship is good to capture the imagination, it has, in spite of a long incubation time since the 1970s, not resulted in more traction than rare words like ailurophile, cereology and discobolus. So how can we come closer to a practical implementation of water stewardship?

It is clear that while water stewardship depends on our scientific-technical knowledge, that alone will not be enough to succeed. It is necessary to move beyond the scientific-technical realm to capture the essence of water stewardship and to reap its benefits.

Ken Wilber (1996) provides a framework to organise one's thinking about the complexity of that question. His idea of a quadrant model is that to capture the essence of a holon or a holistic whole, one has to address four different domains of that whole: the subjective (the I), the intersubjective (the we), the objective (the it) and the inter-objective (the its).

The subjective domain is about each of our individual practices. Here we need to answer questions about our individual role; if we believe it to be relevant (as I do), this is where we train our mindset and improve our connection to our heart and body. This is where we develop necessary competences in the technical as well as in the social domain.

The inter-subjective space is the we-space. This includes our shared stories, our decision systems, the trust we manage to build per Ostrom's recommendation. This is the space of our collective culture, what is considered normal behaviour here? How high is the integrity of our organisation? etc.

The objective domain is the domain of the 'thing' we are working with, in this case 'water'. It can be in the form of our water utility infrastructure, plants and components. It can also take a river, a groundwater reservoir, a water catchment area as the centre of attention.

The inter-objective domain is the domain not only of our central object, but the objects around that are related. If our object is a utility, the inter-objective space is all the stakeholders, who may be organised in an industrial symbiosis – or not. This also includes a number of other key stakeholders, as could be seen in the case about Tisso.

The point of Ken Wilber's model is that we need to address all four quadrants when we want to improve or progress. Together these four aspects can describe everything there is to know about a holon. A holon means something that is a whole as well as a part. Holons are found everywhere and any domain can be viewed as a holon. When considering all four quadrants, we ensure to apply a holistic view of our problem and the odds of success increases drastically (see Figure 20).

<table>
<tr><td>

Subjective – the I space
- Personal mindset and practice of tuning in to the role of water stewardship
- Personal competences and modes of working
- Clear awareness of purpose and meaning
- A sense of shepherdship towards nature and life
- Servant leadership

</td><td>

Objective – the utility space
- The physical structures
- A body of information to gather around
- Relevant tools for analysis
- Internal practice for optimal operation – documented and competence based
- A certification system, e.g. water stewardship or ISO 14001
- The legal framework of its operation

</td></tr>
<tr><td>

Inter-subjective – the we space
- Decision systems
- Utility culture of purpose and culture
- Methods for addressing the emerging future
- A sense of collective integrity
- A body of collective story telling of stewardship role, of nature, of the system the utility is embedded in and how the utility is at service

</td><td>

Inter-objective – the symbiosis space
- Stakeholder collaboration
- Deep ecology understanding of nature and the natural ressources
- Circular economy/industrial symbiosis
- Supply chain interaction based on life cycle analysis
- Transparency of the utility to its surrounding

</td></tr>
</table>

**Figure 20** Example where a utility is considered as a holon. (*Source*: Ingildsen)

*'We did not create the earth, it was given to us. Instead of just "take, make, sell, use and dump", our role is to be good stewards who pass what has been given to us on to the next generation in the same or better condition as the one in which we received it.'*

Otto Scharmer (2013)

**Your reflections:** What parts of your work have elements of water stewardship? Are there foregone opportunities to be a steward of water? What are the main barriers for water stewardship in your life?

# Chapter 5

# Searching differently

It is time to look for significantly different tools to be able to analyse the situation differently. There is something insufficient about the classical linear and simple cause-effect kind of understanding. There is a tired bluntness in our analytical tools.

In my search for new methods and new insights I have joined a four-year course called Sustainable Co-Creation. It is strange looking back at the decision process for this. I was on the verge – literally – on signing up for a Copenhagen Business School three-year MBA (Master of Business Administration) course. There is an admission test for joining and the date for that was set. My CEO had suggested it and it sounded like a natural next step given the career path I was on. But as the date approached, it was as my whole body conspired against me. It was as if my full nervous system sounded all the alarms it could. What kept on reappearing in my mind was something to the effect of 'this is the old world, this is a continuation of the system as it is, this is not mild, kind and caring'. The evening before the test, I called my CEO and told him I couldn't go through with it – I simply didn't believe those skills to be the right skills and the right mindset for a better future.

Perhaps he was puzzled by this decision, but if he was, he didn't sound so. It took him less than seconds to reorient himself. 'OK, if you think so, I trust you are right. So instead you and I have to figure out what kind of leadership training is right for you – you will have to design it yourself'. This was the best message I could imagine.

The 'education' hence mainly consisted of the experiments described in Chapter 2, the writing of this book and participation in the Sustainable

Co-Creation Course. The attraction of the course was especially the teacher Michael Stubberup. A few years earlier I had been given the book 'Heart Prayer' (Stubberup, 2004). The book had made a deep impression on me. It was about the Eastern Christian tradition of heart prayer. A branch of Christianity I didn't know about. The insights from the heart practice developed over more than one thousand years by monk practitioners called hesychasm. The insights were profound and deeply inspirational.

> *'What is a charitable heart? It is a heart which is burning with love for the whole creation, for men, for the birds, for the beasts ... for all creatures. He who has such a heart cannot see or call to mind a creature without his eyes being filled with tears by reason of the immense compassion which seizes his heart; a heart which is softened and can no longer bear to see or learn from others of any suffering, even the smallest pain being inflicted upon a creature.'*
>
> Isaac of Nineveh (Stubberup, 2004)

Professor Steen Hildebrandt is a Danish academic and author on business, organizational theory and management, a household name in management in Denmark. Together these two very different persons have been working for decades to obtain clarity as to what sustainable co-creation means and how it is brought about. Together they authored the book 'Sustainable Leadership: Leadership From the Heart' (Hildebrandt and Stubberup, 2016). At their course we practice methods for systemic change individually as well as in groups. The following description of methods is inspired by their work and our joint developmental work in the course.

In Danish the word 'leading' has the same meaning as searching – and I think it is true that when we search for the new, that is a kind of leadership. Here I will describe two types of searching: searching with organisations and searching within yourself.

## SEARCHING WITH ORGANISATIONS

Theory U is a framework and a set of tools that Otto Scharmer and his team at MIT have developed for the purpose of initiating system change, see for example Scharmer (2013).

His starting point is that as with faultlines along with tectonic plates are places where earthquakes take place, there are similarly fragile faultlines in our collective social body. Especially three faultlines are pressing at this stage of history: (1) our relationship with nature and our planet; (2) our relationship with one another; and (3) our relationship with ourselves.

Faultline one constitutes an ecological divide which for example can be observed in the increasing pressure on water supply and water pollution, it can be seen in the loss of topsoil, it can be seen in the area of climate change, it can be seen in the breakdown of eco habitats and the following irreversible loss of species. This has led to unprecedented environmental destruction and loss of nature. A key figure is that the current global existence of human society uses 1.5 times the regeneration capacity of the earth.

Faultline two is a socioeconomic divide which can be seen in the gap between those who have and those who have not. It can be seen in the gap between the well-fed and the one in eight who go hungry to bed; it can be seen in a world where billions of people still live at levels of poverty where their most basic needs are not met, and it can be seen in inequality of ownership and income. This has led to unprecedented levels of inequity and fragmentation – resulting in the loss of felt society and solidarity. Less than 100 billionaires own as much as half of mankind combined, i.e. more than 3.8 billion people.

Faultline three is a kind of spiritual divide and can be seen in decreasing levels of happiness and increasing levels of suicide, depression and burn-out. It is a divide between self and Self and causes a loss of 'meaning of life'. Every 40 second, someone commits suicide. More than 800,000 people commit suicide per year. This is more than the sum of people who are killed by war, murder and natural disasters combined.

According to Scharmer, the three divides are connected in that the inner void of faultline three causes increasing consumerism, which causes an increase in the ecological divide which again leads to increases in the social divide. There are internal feedback loops between these three divides that seem to increase the severity, and hence the fragility, of the whole collective social body of our global human society. It is us – all of us – who are creating the results, and every day the results are being reproduced worldwide as if by an army of zombies caught in an unbreakable collective story loop.

Scharmer captures the essence of this 'zombie-state' in the idea of 'absencing'. Often we handle difficult situations with our head on auto-pilot. Scharmer calls this absencing, and I can recollect having done this more times than I can count – to avoid the inconvenience or difficulty of doing deeper work. The absencing process follows this process:

(1) 'Denial', i.e. not seeing what is going on, eyes-wide-shut.
(2) 'Desensing', i.e. reducing, not connecting with or lacking empathy with other.
(3) 'Absencing', i.e. losing the connection to one's higher self, one's ideals and corner-stone values.
(4) 'Delusion', i.e. being guided by illusions and lies that one tells oneself to keep on.
(5) 'Destruction', i.e. damaging others and ourselves.

This leads to a kind of tyranny characterised by one truth ideology, an us-versus-them, rigid collectivism and a one-will fanaticism. Another way Scharmer expresses this effect is 'a closed mind, a closed heart and a closed will'. When we handle difficult situations based on absencing, we bring ourselves into a landscape where the danger of the faultlines increases, and so we create new 'difficult situations'.

The central question is then: how do we stop creating results that nobody wants?

Scharmer suggests that we learn to discriminate between two types of learning. One is the classical type of learning, where we learn from the past, the other kind of learning requires us to train our ability to 'learn from the emerging future'. To change our loops we must learn to learn from the emerging future founded on our 'highest potential'. It is a kind of entrepreneurial approach to life, where the participants in the learning process connect to their own source and to their highest future potential.

The issue of water clearly belongs to the ecological divide. However, a key insight is that this divide cannot be repaired unless repairment happens in the social field and in the spiritual field as well. A central statement in Theory U is 'the success of an intervention depends on the interior condition of the intervenor'. This calls for working with our interior. To make a profound change, we have to work from our own self, our own journey of becoming our higher self. The change is profound and has infinite consequences.

So, as the important connection identified by Scharmer starts with the inner void, the faultline between self and capital-S Self, this is where we need to work hardest right now. This is the 'blind spot' where it all emerges – working with the source of what emerges from our human societies, from each of us. And even further back to the source in ourselves. Hence, all work on the collective, like searching with organisations, has to be grounded in the participants not absencing. Change that is good, true and beautiful cannot succeed if we act like absencing zombies.

Theory U is a process that works with what Scharmer calls presencing as a kind of opposite concept to absencing. Scharmer explains the word presencing like this:

*'To sense, tune in, and act from one's highest future potential – the future that depends on us to bring it into being. Presencing blends the words "presence" and "sensing" and works through "seeing from our deepest source".'*

Otto Scharmer (2013)

Entering the state of presencing is not as foreign as it may sound. Most people have experienced this spontaneously in their lives. It happens when we have a moment of surprise, profound insight – when we have an encounter with our deeper levels of awareness. The feeling of being in this 'mode of operation' is profoundly distinct when we experience it. We are invigorated, curious and excited.

The 'new' in Theory U is a comprehension of the process leading to the presencing state for groups, including an understanding of the pre-conditions necessary and the frame that needs to be established around the work for it to work. The miracle is that it is possible to consistently create conditions that allow groups to create collaboratively from these deeper levels of awareness.

Theory U is a framework for this journey and provides a number of tools for each stage of the journey. The process of Theory U is shown in Figure 21

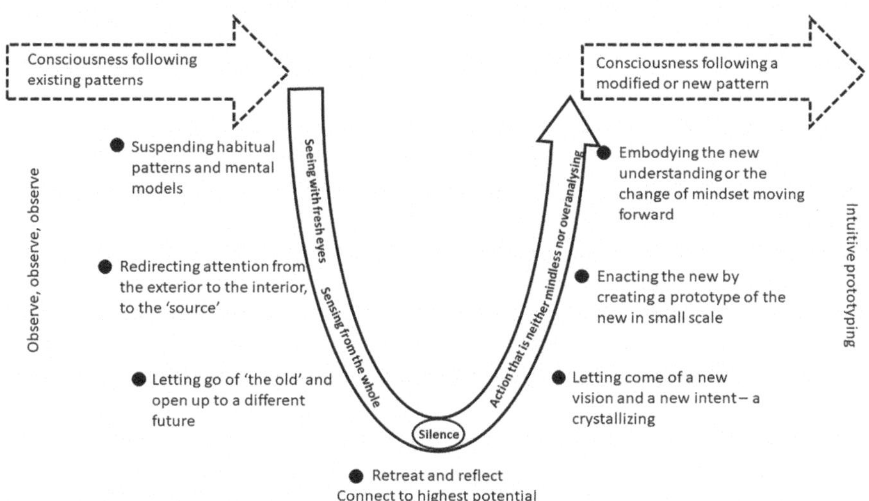


**Figure 21** The process of Theory U according to Otto Charmer. (*Source*: Adapted by the author from Scharmer, 2013)

(I have changed the descriptions to make it more appropriate to engineers, but for a deeper understanding it is a good idea to study Scharmer's own wording (Scharmer, 2013)).

In overview, the process of Theory U is divided into three terrains: the left-hand side, the bottom and the right-hand side. The headline of the left-hand side is 'observe, observe, observe'. In the bottom, there is a retreatment period of reflection, a silent point of transformation. The right-hand side is the path of action characterised by rapid, intuitive prototyping and adapting the prototype until it fits with 'the whole'.

When we go into the details of the process, it consists of five main actions or movements: Co-initiating (left-hand side), Co-sensing (left-hand side), Co-presencing (bottom), Co-creating (right-hand side) and Co-shaping (right-hand side).

The process is co-initiated from the normal state of our consciousness exiting our existing patterns of 'learning from the past'. The co-initiating is about finding common intention, by listening to our own and stakeholders intention and listening to 'what life calls you to work with'. A method for this is to create deep listening in the group, where we learn to understand each other's perspective on the issue at hand. Each participant makes an effort to connect to his/her own intention and what life calls him/her to do. This is shared, and a shared intention is developed around the issue including an understanding of the group's driving force to work on this issue. It also includes identifying critical questions and a list of possible learning journeys that the group could take to get different perspectives on the system they are trying to change. When the group have identified and worded the issues as precisely as possible, they are ready to start the co-sensing part of the Theory U process.

This stage of the process is down the left-hand side of the U and is called co-sensing. Here everything has to do with observing. The process is in fact, called 'observe, observe, observe'. The 'observing' follows three steps. The first step is to suspend habitual patterns and mental models, to observe with an open mind from a new angle. Scharmer suggests 'looking at reality from the edges of the system'. Too often we see things from the echo chamber that we are in, our perception is under an invisible group-think with the people that surround us. So the point is to find new points of observation and look at the chosen issue anew without the baggage of what one already knows. The second stage is to redirect attention from the studied object to the interior, to see oneself in the system, to understand that you are part of the system, to get an 'ecosystem awareness', saying 'I am not me observing the system, I am included in the system, and through me or through us the system is now seeing itself'. The third step in this process is a subtle 'letting go', releasing the mental grip on the system, letting go, in order to allow something new to arrive, to create inner space for it. Letting go without knowing what will arrive.

Another way of explaining the three parts of this co-sensing process is that one has to let go of the inner voice of judgement, i.e. an open mind, one has to let go of the inner voice of cynicism, i.e. an open heart and one has to let go of fear, i.e. an open will. What is intended is that everybody in the group gets into contact with their inner source of knowing.

Based on the observations, the issue to be tackled may be redefined, refined and honed.

As the group arrives at the bottom of the U, the stage of co-presencing comes into focus. At this point, a profound shift of attention is carried out from head to heart. Practically, a silent retreat in nature may be the most conducive conditions for getting in contact with oneself. To resonate over the questions 'Who is my self' and 'What is my work'. This can be carried out both together or one-by-one. The session starts with the group sitting in a circle and sharing reflections, and is followed by a silent retreat, finally the group returns and shares a distilled capture of the essence of each person's insight. The group's sharing sessions are held in a respectful and concentrated atmosphere honouring that something important is about to happen. The sharing of reflections leads to starting points for prototyping in the processes of co-initiating and co-sensing.

If the above two journeys are associated with the head and the heart, the third right-hand journey is associated with the body. In that way, the process has connected the head, heart and body. In this co-creation stage, the group works with prototypes. The prototype need only be 80% ready before being presented to the relevant stakeholders.

The prototype should, however, follow the guidelines: aim for the prototype being relevant, revolutionary, rapid, rough, right, relationally effective and replicable. During the process, a balance should be found to, on the one hand, avoid the dangers of acting too mindlessly fast and on the other hand to avoid acting too slow and potentially falling into 'analysis-paralysis'. When 1–3 prototypes have reached the described level, they are to be presented to stakeholders for feedback. The prototypes are then adapted or changed according to the feedback, preferably with a number of feedback loops.

The purpose of the co-shaping phase is to scale the prototype, making it 'live' in the real world. This important work of reaching fruition for the work has to be done as in a continuation of the spirit that the group has worked until now. The group acts from a shared awareness and a continuous feedback loop between the whole and the parts and this is to be further instituted in the following implementation process. During this work there has to be a continued building of capacity in the group. And new group members have to be welcomed in a way that ensures that they can take a share in both the awareness and the vision.

When I look at the two processes of absencing and presencing, something dawns upon me. The initial feelings I began this book with, the feelings of desecration, banality and apathy, were not new to me. But for a long time I avoided the feelings and dodged them as best possible until I mustered in myself the required open mind, open heart and open will to face these feelings, to venture into an understanding of what they meant and tentatively find ways forward. Hence, this book project is a Theory U journey in itself. I recognise all the states described in this process, but found it to be a much more messy process with lots of 'running back's' and 'getting ahead of myself's'. The process has created a sense of inner peace, courage to look forward and an optimistic readiness for the change ahead.

> **Your reflections:** Can you recognise the difference between presencing and absencing? Can you remember situations in your life where you did one or the other? Do you remember why? Do you recognise it when others do the one or the other? How would you know the difference? What is the difference in sensed experience in the two cases? How can you enable or give space to presencing in your work with others?

## SEARCHING WITHIN YOURSELF

Sustainability sometimes becomes an elusive word, and in all the analysis we lose track of the bigger picture. Sometimes it is good to remember that we ourselves in our bodies have deep experience with sustainability and its opposites. In short:

> *'The explanation draws on the mechanisms of life itself and on the conditions of its regulation, a collection of phenomena that are generally designated by a single word: homeostasis.'*

Antonia Damasio (2018)

Homeostasis is the same concept that drew Lovelock to his insights about the earth as a Gaia-system. Similarly, the body consists of numerous interlocking homeostatic systems are. Homeostasis involves negative feedback loops that ensure that various states keep within sustainable life-enabling ranges. The typical example is how the body regulates its internal temperature to a constant of 37°C (98.6°F).

But homeostasis also works on levels that require conscious cooperation by the person.

> *'Feelings are the mental expressions of homeostasis, while homeostasis, acting under cover of feeling, is the functional thread that links early life-forms to the extraordinary partnership of bodies and nervous systems. That partnership is responsible for the emergence of conscious, feeling minds that are, in turn, responsible for what is most distinctive about humanity: cultures and civilizations.'*
>
> Antonio Damasio (2018)

An important part of my journey of understanding water stewardship has been about understanding myself. The work of Antonia Damasio makes it clear that these two things are connected and why. Most of the homeostatic feeling apparatus work without our paying attention to it. From a 'change point-of-view' this has the unfortunate consequence that we act primarily instinctively on the feeling inputs we receive from the homeostatic system. However, when we make the required effort to make these inputs the object of analysis, quite different solutions present themselves.

Stubberup and Hildebrandt suggest a method of Smallest Possible Change that I have found useful. The concept of the Smallest Possible Change builds on ideas and insights from the theory of life as a self-organising system presented by Varela and Maturana (1980). One purpose is to increase our own ability to understand sustainability by examining our own system, i.e. our own human being. The method teaches us important lessons of what systems sustainability means practically. While the theory behind it is complex, the practical method is simple and self-organising towards incremental complexity and effect. This means that the initial usage is simple, but as it is repeated over time it causes an ever-deepening understanding of our sustainability.

There may be different nuances to how to apply the method and one is free to adapt to one's ideas and preferences. The way I apply it is rooted in my need for a disciplined habit. By applying the method on a daily basis I ensure it becomes a habit that helps me ensure my continued learning process.

Each day consist of positive and negative events, i.e. good things happen (attraction), and bad things (repulsion) happen. We have nice emotions and difficult emotions, some people are nice to us, some are not.

The first step is to make a daily recording of the two most positive events and the two most negative events. I do this in the evening just before I go to sleep. My experience is that even if you haven't thought of it that way during the day our 'human-being system', or our nervous system, will answer the question relatively effortlessly, and the same is the case when you ask for the most negative events. This is a small and enjoyable thing to do and is continued until it constitutes a routine. It resembles the general advice of 'gratitude journals', but I prefer this method because it recognises both the positives and the negatives, so it does not deny the negative of our experience.

When the above first step works well and it has become part of a routine, the method is extended. The second step is to 'relive' the two-by-two positives and negatives by recalling the emotional feeling in your body of each event for a short while. The purpose of doing this is to get a sense of the full spectrum of emotions of the day. And by acknowledging and re-sensing these maximum positives and negatives, you enable yourself to take a step back and sense a kind of new neutrality from which you can move on.

Over time, this strengthens what is called the 'witness-function'. The 'witness-function' is an ability in your consciousness to distance yourself slightly from what you are doing. You are able to witness your own mental processes. This allows for a small gap between the input you get and the response to give; a time gap that allows you to gauge your feelings and come up with an appropriate response to the situation. So instead of *being* angry, happy, sad etc. you have a slightly changed stance of rather *experiencing* anger, happiness or sadness. This is a subtle change in perspective, where you do not perceive yourself as the emotion, but feel yourself feeling the emotion. It is a change that, as it is trained, improves your ability to observe your own system, including your nervous system and cognitive processes.

My first experience with this was a small registration through the day of – 'ah this will probably be one of my positives' (or negatives). Other effects of the witness function are a sense of time being stretched. When the witness function is strong, a day feels much longer in a way where boredom is not a part of the experience. This stretching of time provides a change in the perceived opportunities for how to act, especially in difficult situations. There is simply more time to think, sense and come up with an appropriate reaction compared to 'made-in-the-moment' automatic reactions. But perhaps even more positively, when the sun is really shining in a nice way or the teamwork is flying, I notice it and feel the joy of it more deeply.

The third extending step is to train naming feelings. For each positive and negative event I try to describe the feeling – just in a few words. It was surprisingly difficult to do in the beginning, trying to come up with a sentence that in a few words captures the essence of the felt feeling. This is a wonderful way to develop a language you can 'speak with yourself' in. And it improves the emotional agility, your ability to explain to others what is going on in you as well as your capacity for 'articulated' empathy, i.e. the type of empathy where you can actually help others understand their feelings.

The fourth extension is to add an intention of a 'Smallest Possible Change' for tomorrow. The idea is to promise yourself to change a small thing tomorrow.

The most important reason for the change to be small is that it must not trigger any alarm signals in your nervous defence system. It should really be something that is easy and nice to do. For the first many experiments with this method, make the change be about self-care. So try in a 'homeostasis-promoting' way to listen to what your nervous system is longing for, to ensure that it thrives, feels safe and is nourished.

At the onset take slow steps to get some experience with the method. A special purpose is to gain trust in yourself, the best kind of self-confidence. A self-confidence that you can trust yourself, i.e. if you promise yourself to change or do something – do it. If you are in doubt of whether you can go through with an act, reduce it or simply don't promise yourself to do it. When you have repeated this process of promising and delivering 10 or 20 times your 'self-confidence' is increased and you may add the load in your 'smallest possible change for tomorrow' promises, but carefully avoid triggering fear responses.

To me this has proven to be a very flexible and adaptive way of improving my sustainability – all the time adjusting to what is happening around me and in me.

If you would like to join similar courses, it may be a good idea to look at MIT U-lab for a start. They offer free online courses at edx.org.

> Your reflections: Are you able to keep your word, when you promise yourself something? Do you have that kind of self-confidence? What if the promise was about self-care first?

# *Chapter 6*
# Blind spots

---

The work of understanding water stewardship is a long identification of blind spots. This chapter reveals a few additional blind spots.

## THE INEFFICIENCY OF 'THE POLLUTER PAYS'-PRINCIPLE

'The polluter pays' was a revolutionary idea when it came up. However, today, we can see that the changed mindset it signifies is insufficiently radical. Nature takes neither cash nor credit cards. Besides, in the current state of social, economic inequality, there is nothing inherently equitable in the richer being allowed to pollute and destroy, while the poorer should be more careful with their actions and effects. 'Pollution is not allowed' ought to be the mantra just like you are not allowed to, for example, steal from other people.

In any field of knowledge there is a hard core of basic assumptions with a protecting layer around that protects the hard core from being questioned, so that the people in that field can continue their work undisturbed. In economy, the basic assumptions have to do with who or rather what generates value. There is a strong orientation towards seeing the entrepreneur as the creator and therefore also seeing the entrepreneur as the main benefactor and decision-maker around the generated wealth. But it makes it hard to see that no entrepreneur creates wealth on his own. The key helpers include society and its infrastructure to help protect his property, roads to transport raw material and goods, electric and water utilities to provide easily available resources. The involved employees are another

© IWA Publishing 2020. Water Stewardship
Author: Pernille Ingildsen
doi: 10.2166/9781789060331_0149

crucial element. And not only them but all the things behind them, ensuring they can meet at work in the morning, well-rested, well-fed, educated and with their kid's taken care of. And it requires nature providing all the natural resources going into the production on a myriad of levels.

Similarly, we should investigate the basic protected assumptions of water and wastewater utilities. First of all, it is elemental to understand that the raw material of water that utilities sell are not paid for. The water is taken out of nature's resource bank for free. The utility's contribution to the value chain is extraction, treatment, distribution and administration. Similarly, on the wastewater part, the load of the effluent is 'paid' by nature, for which there is, of course, no compensation. This fundamental system is administered by authorities by means of permits to extract or to pollute.

These core premises easily hide from our view. We do not experience that we are consuming while externalising the negative consequences. By their payment, water consumers feel free of the inner sense of externalised consequences. They have paid for their water and wastewater and as such are free of guilt and debt of conscience – they are simply not aware that they only paid for the utility's processes.

We must be alert to this process of blind externalisation of consequences. For example, when we work with the sustainable development goals, it may soothe such feelings of guilt. But a closer study (Independent Group of Scientists appointed by the Secretary-General, 2019) reveals that several indicators are moving in the wrong direction, and others are moving too slowly to reach their destination in time (2030). The report reveals that no country is currently able to meet basic human needs within the biophysical boundaries. Those who are able to meet basic needs have the greatest trespassing of biophysical needs and vice versa – some are neither able to meet needs or stay within boundaries.

We may be unwilling to scrutinize these things out of fear of having to sacrifice various comforts. However, when we look closer, we see that we are already sacrificing a lot of things, but they are mostly invisible to us. In his poem 'Questionnaire', Wendel Berry tries to make the suffering of our actions visible. He asks: how much poison are you willing to eat for the sustaining the current system? How much evil are you willing to do? The questions continue when it comes to water. When working with water, we are in the frontlines of these questions, as we answer them on behalf of everybody using our products – which often is most of the community in which we operate. Thinking only in terms of whether a thing is legal or not or whether it is common practice or not, may not suffice moving forward. We have to also find that our actions and production methods are in harmony with our inner sense of water stewardship. Detecting these incongruencies may take time, finding new solutions take time, hence perseverance and patience is required. However, never have so many people shared a joint vision of sustainability than now – and inspiration is everywhere to be found – hence it is in many cases a question of taking up the mantle.

> **Your reflections:** What practices do you see that you wish to change? Where do you take up the mantle of improving water practices?

## THE HYDROLOGICAL CYCLE IS MORE COMPLEX THAN WE IMAGINE

In 2018, Professor Malin Falkenmark of Stockholm Resiliency Institute won the Blue Planet Prize for her work in hydrology (Falkenmark, 2018). Falkenmark calls the hydrological cycle the bloodstream of the biosphere or the life support system. We often, in our thoughts, simplify the water cycle into a simple round water circle. But, as Falkenmark points out, reality is a bit more complicated (Falkenmark *et al.*, 2019) as can be seen in Figure 22.

Via her accounting of how the bloodstream flows she has identified two water streams circles, the blue water stream, which is the part of precipitation that ends up in rivers, lakes and groundwater and eventually flows via surface run off to the ocean. This constitutes, on average, 40% of the precipitated water. The other is the green water which via complicated routes through the root zones and biomass ends up in the atmosphere again. The two types of water together have eight main functions; the green water functions are:

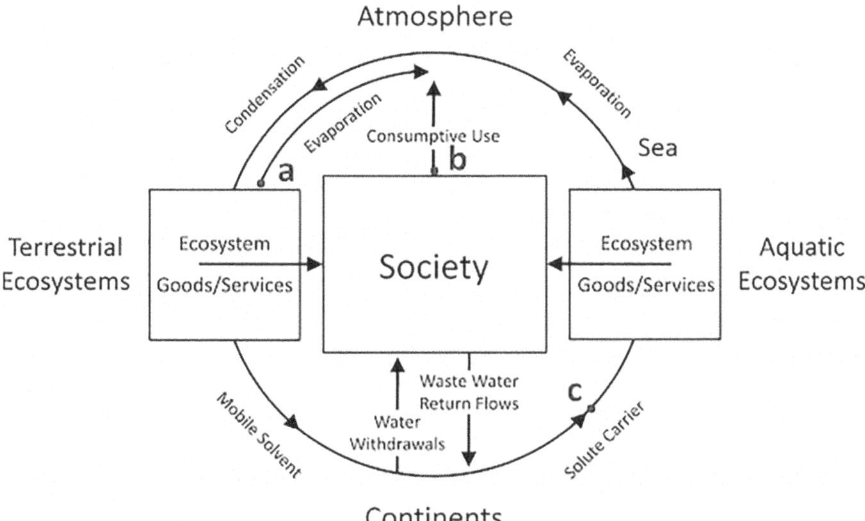


**Figure 22** The hydrological water cycle according to Malin Falkenmark. (*Source*: Rockström *et al.*, 2012)

(1)  Generation of air moisture which works as a green-house gas – the most potent greenhouse gas in the atmosphere, responsible for 20–25% of global warming. Without water, the planet would have been 30°C colder than it is.

(2)  Biomass production, where plants collect water from different levels of the root zone in the soil.

(3)  Moisture feedback to the atmosphere, where water is recycled over land.

The blue water has five functions:

(1)  Supplying society with water for its various needs.

(2)  Carrier of nutrients and pollutants throughout the system. Flushing pollutants from land to the ocean.

(3)  Upholding the aquatic state in rivers and lakes and thus working as water storage.

(4)  Production for agriculture in the form of irrigation.

(5)  Aquatic biomass growth.

Falkenmark points out that human interaction with the hydrological cycle disturbs it, not only by the extracted water for urban, industrial and irrigation uses but also from land use. It is disturbed in terms of flows and in terms of water quality as pollution is led to the environment, but it is also disturbed through the eco-systems that are performing 'eco-system services'. Green water degradation includes desertification, savannization and salinization, while blue water degradation includes basin closure, aquifer depletion, eutrophication and aquatic systems collapse.

*'The water-resource challenge of the future is more complex than previously portrayed – it is not only a question of water allocation among irrigation, industry, and municipalities but involves difficult decisions for balancing green and blue water for food, nature, and society. It will change the role of water-resource planners and managers. Water resources planning and management will have to incorporate land-use activities consuming green water and its interaction with blue water, generating surface runoff and groundwater recharge.'*

Falkenmark and Rockström (2006)

---

**Your reflections:** What are the dangers of simplification?

---

## THE SUSTAINABILITY CRISIS IS ALREADY HERE

For decades, I have feared 'The Sustainability Crisis of the Future'. There was a sense of emergency and alarm, the need to act to avoid a future catastrophe. But today, the reality is that the catastrophe is already here – moving forward in its silent pace. Animal species have gone extinct and will never come back, and the green-house effect already has devasting effects in the Arctic and Antarctic regions where glaciers melt away and this year (2019) in Australia where bush fires reached new records. The biodiversity of insects has been reduced substantially. The only native forest in Europe lies at the border between Poland and Belarus; all other old forests are gone. Temperature and rainfall patterns are changing, nutrient overload in the oceans are changing the living conditions of the ocean. The sound of nature's silence is broken in so many places on a permanent basis above and below sea level. Residual medical substances are changing life in the oceans. Islands of plastics the size of continents are floating in the oceans. Whales with their stomachs full of plastics wash-up on the shores. Megacities around the world experience major water supply emergency situations.

To acknowledge that the sustainability crisis is already here is ignition to action. The crisis is not a hypothetical threat out there in the future. It is unfolding with slow but great strength right now. Reaching the SDGs is not the end of the crises, it is the beginning of a turning of the tide. And even when the tide starts turning it still holds great momentum for further destruction.

To me, to acknowledge the sustainability crisis is happening now releases tensions and provides energy and determination.

---

**Your reflections:** Where do you see signs of the sustainability crisis being here already? How can you see it in your local area? Can you see it in yourself?

# *Chapter 7*

# Utopian vision

It is curious to observe that most science fiction novels are dystopic in nature. These stories may bias our imagination toward the more sinister side. However, in Charlotte Perkins Gilman's novel Herland (1915), an example of an ecological utopia is provided. The story is told from the perspectie of three explorers who discover a country the size of the Netherlands that has been isolated for 2000 years from the rest of the world. In this country there are only women who, by some process, have learned to give birth without the involvement of men. The country has been cut off in the mountain due to an eruption of a volcano at the time of war when the men were warring outside the country. First, the remaining women considered suicide but decided to live their lives to the end. At some point, a woman conceived a child and 'the whole spirit of the country changed from mourning and mere courageous resignation to proud joy'. This became the foundation of a new race of women able to regenerate the species.

The story is about how a utopian society might develop when the culture is focused on 'mothering' with the continuous and conscious development of a country toward a better place for people to grow up in. The novel is from 1915 and is a comment in a sense from the suffragette movement. But, putting that aside for a moment, it describes what it would take to create a society of increasing beauty and happiness of and for its inhabitants.

A few quotes from the book give a sense of a completely different possible future:

*'Yet, they were not old women. Each was in the full bloom of rosy health, erect, serene, standing sure-footed and light as any pugilist.*

© IWA Publishing 2020. Water Stewardship
Author: Pernille Ingildsen
doi: 10.2166/9781789060331_0155

> *Never, anywhere before, had I seen women of precisely this quality. Fishwives and market women might show similar strength, but it was coarse and powerful. College professors, teachers, writers – many women showed similar intelligence but often wore a strained nervous look, while these were as calm as cows, for all their evident intellect.*
>
> *... one of the things most impressive about them all was the absence of irritability.*
>
> *All the scraps and leavings of their food, plant waste from lumber work or textile industry, all the solids combined – everything which came from the earth went back to it. The practical result was like that in any healthy forest; an increasingly valuable soil was being built, instead of the progressive impoverishment so often seen in the rest of the world.*
>
> *They had the evenest of tempers, the most perfect patience and good nature – one of the most impressive about them all was the absence of irritability.*
>
> *There was something to their criticism. The years of pioneering lay far behind them. Theirs was a civilization in which the initial difficulties had long since been overcome. The untroubled peace, the unmeasured plenty, the steady health, the large goodwill and smooth management which ordered everything, left nothing to overcome.*
>
> *They found themselves in an immediate environment, which was agreeable and interesting, and before them stretched the years of learning and discovery, the fascinating, endless process of education.*
>
> *' "Have you no respect for the past? For what was thought and believed by your foremothers?" "Why no," she said. "Why should we? They are all gone. They knew less than we do. If we are not beyond them, we are unworthy of them – and unworthy of our children who must go beyond us." '*

Charlotte Perkins Gilman (1915)

The three explorers are at first 'imprisoned' in the country. But it is not a normal prison; it is a very nice place, clean and surrounded by teachers; teachers who are eager to learn about the world outside the boundaries of Herland and eager to teach the language, history and philosophy of Herland. The novel is a description of this learning process and contains a comparison between the two places, Herland and outside of Herland, of every aspect of society from handling of criminals, health and illness, child-rearing, food, clothes, poverty, strife, religion, marriage, etc.

This leads to an increasing embarrassment of the three explorers, who were actually quite proud of their American homeland at first. But with this country of wise, selfless people, it is quite clear how many opportunities of 2000 years of development have been foregone, how much a country could have developed, had it not been for the recurring wars, strife and conflict, in essence by its lack of ability to control the fear and hatred. These emotions were maybe also stirred at some point in the Herland race, but the highest effort is made to ensure a safe and nourishing environment for the children and great energy is put into improving the children's learning processes. The aim is that learning is not felt like a difficult obstacle that children have to overcome, but rather as an explorative and playful joyful activity of childhood. By this 'process' of always improving everything and not holding old beliefs in unquestionable esteem, the country develops as they overcome everything that seems to counter their increasing sense of morality and community. The people and the society described are in essence healthy, serene and in strong integrity with the prevailing 'theory of life'.

As to the relationship to nature and the environment:

*'Here was evidently a people highly skilled, efficient, caring for their country as a florist cares for his costliest orchids.*

*… a land in a state of perfect cultivation, where even the forests looked as if they were cared for; a land that looked like an enormous park, only it was even more evidently an enormous garden.*

*All we found moving in those woods, as we started through them, were birds, some gorgeous, some musical, all so tame that it seemed almost to contradict our theory of cultivation – at least until we came upon occasional little glades, where carved stone seats and tables stood in the shade beside clear fountains, with shallow birdbaths always added.*

*The practical result was like that in any healthy forest; an increasingly valuable soil was being built, instead of the progressive impoverishment so often seen in the rest of the world.*

*They loved their country because it was their nursery, playground, and workshop – theirs and their children's. They were proud of it as a workshop, proud of their record of ever-increasing efficiency; they had made a pleasant garden of it, very practical little heaven; but most of all they valued it – and here it is hard for us to understand them – as a cultural environment for their children.'*

Charlotte Perkins Gilman (1915)

After reading the story, it also exists a bit in the mind of the wondering reader. Wondering if this could have been a possibility, wondering if we have wasted 2000 years of opportunity for much greater progress, wondering what stands in the way for really applying similar continuous and conscious processes of dedicated mild and balanced improvement. The novel raises questions about some of the basic premises of our world mindsets, so basic that we are surprised when having them pointed out. The reader wonders how easily we have accepted detrimental concepts as 'natural laws' when they are really not. A thing that made me laugh was her comment: 'The tradition of men as guardians and protectors [of women] had quite died out. These stalwart virgins had no men to fear and therefore no need of protection'.

What Charlotte Perkins Gilman is describing is not so much a country without men, as it is a country that has not been driven by egoic principles. A country where a dedicated communal mindset has worked for generations on improving the process of growth of people from birth and onwards. It is about nurturing an inquiring mind, a free mind that works for the joy of it and for further improvement.

In Herland, the inhabitants take care of nature as they are part of it. Their planning and doing always have the benefit of the long term in mind. The long term of generations ahead. In Herland, they have the expression of 'loving up' – as in contrast to 'loving down'. The concept of motherhood is described as:

> *'I don't mean the underflannels-and-doughnuts mother, the fussy person that waits on you and spoils you and doesn't really know you; I mean the feeling that a very little child would have, who had been lost – forever so long. It was a sense of getting home; of being clean and rested; of safety and yet freedom; of love that was always there, warm like sunshine in May, not hot like a stove or a featherbed – a love that didn't irritate and didn't smother.*
>
> *There you have it. You see, they were Mothers, not in our sense of helpless involuntary fecundity, forced to fill and overfill the land, every land, and then see their children suffer, sin, and die, fighting horribly with one another; but in the sense of Conscious Makers of People. Mother-love with them was not a brute passion, "a mere instinct", a wholly personal feeling; it was – a religion.*
>
> *To them, the country was a unit – it was theirs. They themselves were a unit, a conscious group; they thought in terms of the community. As such, their time-sense was not limited to the hopes and ambitions of an individual life. Therefore, they habitually considered and carried out plans for improvement which might cover centuries.*
>
> *They applied their minds to the thought of God and worked out the theory that such an inner power demanded outward expression. They lived as if God was real and at work within them.*
>
> *Every woman of them placed motherhood not only higher than other duties but so far higher there were no other duties, one might almost say.'*
>
> Charlotte Perkins Gilman (1915)

It is clear from my work with water stewardship that in the society we have built over the last hundreds of years, we have come to a point in history where it is necessary to think long and hard about water stewardship. When swimming in the ocean, steadily working to keeping my head over water, or standing in a fiord where the waves try to overthrow me or when I see the miracle of water added to the garden after a long dry summer, I wonder: 'am I going to be a steward for water?' Water has its own powers that are much greater than mine. Hence, water stewardship is not a 'cuddly nursing kind of care'. When understood from a mothering perspective, as Gilman suggest, things get clearer. The stewardship is for all living, and an important guiding principle is to make a world for our children. Not just *my* children, but for all children on Planet Earth now and moving forward. From that point of view, water stewardship is about respect, humility and thankfulness.

---

**Your reflections:** What does your imaginary utopia look like?

# Chapter 8
## About the act of visioning

In her speech 'Down to earth' from 1993, Donella Meadows starts by reminding her audience that there is a lot of work to be done in implementing the multitude of steps necessary to move into a sustainable future (Meadows, 1993). For that we need a lot of talented people, implementers, who can make things happen, who can write policy, acquire resources, make the resources useful and create new things and find new ways.

Behind them we need models, data and theory that can guide us in taking the right steps to take us where we want to go. People who understand how we got to this point, who knows what went wrong and what went right. People who can find their way forward based on data and analysis.

And then behind those we need people who can provide vision, i.e. clarity about the goals and themes of the future; goals and themes that are articulated and can be discussed among us, so that all can participate, contribute to and share that vision.

When it comes to creating a sustainable world, the deficiencies in all of the three arenas are evident. However, she states, we as humans have a tendency, when problems arise due to these deficiencies, to focus primarily on the implementation level – we want to 'go out and do something about it'. The problem with that approach is that before we act we need to be sure that our models and data are correct, or we might – as we have seen so often – act on a false foundation. And before we even get our data right, we need to know which data are relevant, i.e. we need to understand where we want to go. We need vision.

© IWA Publishing 2020. Water Stewardship
Author: Pernille Ingildsen
doi: 10.2166/9781789060331_0159

Donella Meadows' professional background is in the model and data part and her first impulse for the speech was to talk about the results of some of her research. She would have made a speech on how to change the economic principles and how we could apply technology better. However, she argues that what we are doing in our problem-solving eagerness is a 'skipping over' of the visioning step – as if we take it for granted.

Therefore, instead of giving the classical speech of models and computer outputs she chooses to speak of visioning and encourages everybody to work with that dialogue. She states that she is completely unqualified for that kind of talk, because she has no training in visioning or communicating vision. But we are all untrained. But we ought to try anyway – even when it goes beyond our classical academic training. The process of visioning is basic to all of us, we can all do it.

Environmentalists often seem to propose non-inspirational visions of sacrifice, frugality, restriction of freedom and increased control. However, if we instead spend time imagining and visioning, something else will emerge. If we instead try to imagine how the world would look like if its environment was healthy; if all water streams were clean and flowing and providing a good habitat for people and nature alike we would get somewhere different in our dialogue. It is interesting how little amount of time we spend to imagine how it would be like to be in such a world. Instead we are occupied with the problems all the time – I wonder if we even take time off to enjoy our small successes when we do succeed?

What is underneath this dismissal of time spent for visioning a brighter future? Donella Meadows has worked with this and the essence of the answers she has heard again and again: 'I can't stand the pain of looking at the world I really want, when I know about the condition the world is in. I can't stand that tension. I really don't want to look at it. I can't share my vision with all of you because I don't know you well enough yet.' Why is it not socially permitted to share our visions and dreams? What dangers hide there? We are fast to deem dreamers naive, idealistic and unrealistic. This is so ingrained in our culture that it is almost an automatic reaction – I can even hear myself judging myself on this.

Donella Meadows' experience, when she found a way to speak with people about her and their vision, was a great relief and the pouring forth of beautiful visions. The words of these conversations would have given a different picture than the current discourse of sacrifice, loss of freedom and increased control. People became invigorated and ready to work for the cause. She explains: 'I have learned how more to tap the part of me – I don't know how to describe it – from which vision comes. It's not the intellect, it's not the mind, it's not the rational set of skills that we have as human beings. It comes from another place, which you may call heart or soul some combination of the two.'

Donella Meadows has implemented a practice of always taking time off before starting a project to vision 'what it would be like if it was perfect'. She spends so much time on it that she knows the vision intimately and can refer to it, when she feels like losing her way. The point, she explains, is to state as precisely as possible what you really want – not what you think you can get. We have to set aside our mental models of what we believe is possible, because what we earlier thought was impossible happens all the time. Secondly, you are not obligated to know or show a way to get to that vision. Such a requirement again puts a spotlight on what we allow ourselves to imagine – also remember that when you listen to others vision.

'My experience', Donella Meadows explains, 'in having now many times created a vision and then actually brought it in some form into being; my experience is that I never know at the beginning how to get there. But as I articulate the vision, put it out, share it with other people and it gets more polished and more real the path reveals itself. And it would never reveal itself if I were not putting out the vision of what I really want and finding that other people really want it too. Holding on the vision reveals the path and there is no need to judge the vision by whether the path is apparent.'

> **Your reflections:** Have you set time aside to work on your visioning in your current projects or your job? How does the world you would love to live in look like? Donella claims that having such an inner vision alive inside prevents her from selling out to something less. Vision gives you the power to stay on course – can you recognize that in you? How can visioning become a practicality for you?

# *Chapter 9*

# Facing the wicked problem

---

We need to face that we live on a spaceship that is operating outside its bounds; it is not sustainable. The most obvious measure of unsustainability these days is $CO_2$-content in the atmosphere. But it is too shallow thinking to believe that by cutting $CO_2$ emissions we will be back on track and can continue with business as usual. The planet and the human culture on it are unbalanced in many and accumulating ways.

Our predicament is what has been named a 'wicked problem' because of complex interdependencies and because one or a few people cannot – even with all their might and wealth – solve the problem alone. I often picture the solution as the view you get in a kaleidoscope: When the pattern changes everything in the pattern changes at the same time. Of course, in the kaleidoscope the mirrors are the key to the well-timed change in the visual patterns in the kaleidoscope, so there it is not so difficult to perform.

To make such a kaleidoscopic transformation throughout every human institution is something else. It is as if every institution stands in the way for everyone else. But of course, what we can do is change a bit, so that the rest of the pattern can change a bit, which will unlock possibilities for new movements. Step-by-step we can change everything. Of course, this can work out; it has worked out before. After all, our human landscape changes all the time.

I believe that my role as a water professional or water worker is to transform into a water steward – step by step – to take on this greater responsibility as best I can. By moving like this we can begin solving the water sustainability problem – small step

© IWA Publishing 2020. Water Stewardship
Author: Pernille Ingildsen
doi: 10.2166/9781789060331_0163

by step synchronised with the global kaleidoscopic changes. We can move, we can lead, and we can follow – everywhere, in all functions and on all scales. Everybody working with water can make a difference in their position – if only by being awake to what is happening, if staying focused on learning and understanding more profoundly, if only visionary enough to be able to act when the opportunity is there. Water heroes come in all sizes. But regardless the change requires something from us.

Everybody working with water – in every position – can make a difference by being awake to what is happening, by staying focused on learning and by understanding more profoundly. Everybody needs to be sufficiently visionary to act when the opportunity is there.

*'Water is the driving force of all nature.'*

Leonardo Da Vinci

---

**Your reflections:** What does water stewardship mean to you? What is important to you? What will be your legacy?

# *Chapter 10*
# I am a water steward

*'I used to think the top environmental problems were biodiversity loss, ecosystem collapse and climate change. I thought that with 30 years of good science we could address those problems. But I was wrong. The top environmental problems are selfishness, greed and apathy... and to deal with those we need a spiritual and cultural transformation and we scientists don't know how to do that.'*

Gus Speth

Having crossed the line for unsustainability some 50 years ago, we continue to move in the wrong direction on the total global bottom line. This is not to say that there has been no progress, there has. But it is as if the destructive forces are stronger than the restorative effects.

I think Gus Speth is right. We, as scientists don't know how to do that. But we as human beings do know how to create a spiritual and cultural transformation. We can change and transform as we have done for generations. The stairways of maturity are mapped ahead; we need to create in ourselves and in our hearts whatever it means to transcend Graves' first-tier stages and go to the fearless, generous second-tier stages. At all times there will be people at all stages of maturity, so we must invent ways to guide them forward without falling too heavily into the shadow sides of each stage. In all our experiments there will be frustrations and room for improvements, but that is not a reason to despair, that is a reason to keep moving forward.

Standing here at a crossway between a vision of a better world and a nightmare of a collapsing unsustainable earth, I wonder if we have two ways of doing it. Either we can peacefully, voluntarily and pro-actively move towards a better and more just world or we will have to 'save the world' with forced hands, in panic and with significant losses in our wake. So many trends point at our human proclivity to

© IWA Publishing 2020. Water Stewardship
Author: Pernille Ingildsen
doi: 10.2166/9781789060331_0165

the latter road. But look again: so many people want to move via the first way, momentum is building. They just don't know how to do it. We need to acknowledge the difficulty and jointly invent the road ahead.

Sometimes I still catch myself at a loss in all this. But I have promised myself that when I get overwhelmed with apathy and only see desecration and banality around me, I will take a deep breath, close my eyes and connect to my heart of hearts. And I will remember that I am a water steward because I choose to be.

---

A water steward is a warrior who does not give up.
A water steward is a poet, who sees possibilities for beauty.
A water steward is a deep ecologist, who continues to learn about nature and its needs.
A water steward thinks seven generations ahead and acts in that faith.
A water steward understands the power of many small steps.
A water steward is an experimenter always learning by doing.
A water steward reconnects where connections have been severed.
A water steward is a diplomat, who negotiates needs and solutions.
A water steward knows technologies and analytical methods that can solve difficult water problems.
A water steward keeps integrating and improving the water system for all.
A water steward keeps working for a greater vision – regardless of the difficulties.
A water steward understands natural law.
A water steward helps and supports fellow water stewards.
A water steward is part of a water fellowship.
A water steward is humble, respectful and thankful when it comes to water.
A water steward stands in full height.
A water steward understands how deeply intimately we are connected with water.
A water steward understands the world as 'a Gaia world' and understands the role of water there.
A water steward will eventually see a new world emerge.

---

*'The difference between the possible and impossible is that the impossible takes longer time.'*

Fridtjof Nansen

# References

Andersen H. C. (1843). The Nightingale (Nattergalen). C. A. Reitzel, Copenhagen, Denmark.

AWS (2019a). AWS standard. Available from: https://a4ws.org/the-aws-standard-2-0/ (accessed 23 November 2019).

AWS (2019b). The AWS standard 2.0. Available from: https://a4ws.org/the-aws-standard-2-0/ (accessed 29 December 2019).

Ballowe J. (2018). Banality's evil: an interview with Ellizabeth Minnich. *Minding Nature Journal*, **11**(1). Center for Humans and Nature. Available from: www.humansandnature.org/banalitys-evil-an-interview-with-elizabeth-minnich (accessed 23 November 2019).

Beck D. E. and Cowan C. C. (2005). Spiral Dynamics: Mastering Values, Leadership and Change. Wiley-Blackwell, Hoboken, NJ, USA.

Benyus J. M. (2002). Biomimicry: Innovation Inspired by Nature. Harper Perennial, New York, USA.

Berry W. (2005). Given. Larkspur Press, Monterey, KY, USA.

Berry W. (2010). Questionnaire from Money and Morals after the Crash. Yale University, New Haven, CT, USA. Also available from: https://reflections.yale.edu/article/money-and-morals-after-crash/questionnaire

Blixen K. and Dinesen I. (1937). Out of Africa. Putnam, New York, USA.

Brundtland G. H. (1987). Our Common Future. Oxford University Press, Oxford, UK.

Campbell, J. (1949) A hero with a thousand faces. New World Library, Novato, CA, USA.

Capra F. and Luisi P. L. (2014). The Systems View of Life: A Unifying Vision. Cambridge University Press, Cambridge, UK.

Carse J. (1986). Finite and Infinite Games. Free Press, New York, USA.

Charlotte P. G. (1915). Herland. Dover Publications, London, UK.

Chatwin B. (1987). The Songlines. Franklin Press, Philadelphia, USA.

Childre D. L., Martin H. and Beech D. (1999). The Heartmath Solution. HarperOne, San Francisco, USA.

Climate Home News (2019). Denmark adopts climate law to cut emissions 70% by 2030. Available from: www.climatechangenews.com/2019/12/06/denmark-adopts-climate-law-cut-emissions-70-2030/ (accessed 23 November 2019).

Cunningham L. (2013). Seven Sisters. Sun Dogs Creations, Plano, TX, USA.

Damasio A. (2018). The Strange Order of Things. Vintage, London, UK.

Dawkins R. (1976). The Selfish Gene. Oxford University Press, Oxford, UK.

DGNB (2019). The DGNB System: Global Benchmark For Sustainability. Available from: www.dgnb-system.de/en/ (accessed 23 November 2019).

Dictionary.com. Definition of stewardship. Available from: https://www.dictionary.com/browse/stewardship (accessed 23 November 2019).

Eirera A. (2012). Aluna. Sunstone Films, London, UK.

Eisenstein C. (2015). The More Beautiful World Our Hearts Know is Possible. North Atlantic Books, Berkeley, CA, USA.

Emoto M. (2004). The Hidden Messages in Water. Simon and Schuster, New York, USA.

Ensler E. (2014). Suddenly, My Body. Available from: www.ted.com/talks/eve_ensler (accessed 23 November 2019).

Etymonline. Definition of steward. Available from: https://www.etymonline.com/word/steward (accessed 23 November 2019).

Falkenmark M. (2018). Shift in Water Thinking, Blue Planet Prize 2018. Available from: www.youtube.com/watch?v=hL_GA9rlYdc (accessed 29 December 2019).

Falkenmark M. and Rockström J. (2006). The New Blue and Green Water Paradigm: Breaking New Ground for Water Resources. Planning and Management. Available from: https://ascelibrary.org/doi/10.1061/%28ASCE%290733-9496%282006%29132%3A3%28129%29 (accessed 29 December 2019).

Falkenmark M., Wang-Erlandsson L. and Rockström J. (2019). Understanding of water resilience in the Anthropocene. *Journal of Hydrology*, **X2**, 100009. Available from: https://doi.org/10.1016/j.hydroa.2018.100009 (accessed 28 March 2020).

Fuller B. (1969). Operating Manual for Spaceship Earth. Lars Müller Publishers, Zurich, Switzerland.

Global Footprint Network (2019). Earth Overshoot Day. Available from: www.overshootday.org/newsroom/past-earth-overshoot-days/ (accessed 23 November 2019).

Gormezano D., Protti T. and Cowie S. (2016). The Marianna Disaster. A Journey Through Brazil's Worst Environmental Disaster. Available from: http://webdoc.france24.com/brazil-dam-mining-disaster-mariana/ (accessed 23 November 2019).

Graves C. (1971). Levels of existence: an open system theory of values. *Journal of Humanistic Psychology*, **10**(2), 131–55.

GWP (2000). Integrated Water Resources Management TAC Background Papers. Available from: www.gwp.org/globalassets/global/toolbox/publications/background-papers/04-integrated-water-resources-management-2000-english.pdf (accessed 29 December 2019).

GWP, INBO (2009). A Handbook for Integrated Water Resources Management in Basins. Available from: www.inbo-news.org/IMG/pdf/GWP-INBOHandbookForIWRMinBasins.pdf (accessed 29 December 2019).

Hardin G. (1968). The tragedy of the commons. *Science*, **162**(3859), 1243–1248.

Hassing J., Ipsen N., Clausen J. T., Larsen H. and Lindgaard-Jorgensen P. (2009). Integrated Water Resources Management in Action. UNEP, USA.

Hildebrandt S. and Stubberup M. (2012). Sustainable Leadership: Leadership from the Heart. Copenhagen Press, Denmark.

IISD (2001). Summary of the International Conference on Freshwater: 3–7 December 2001. Available from: https://enb.iisd.org/download/pdf/sd/sdvol66num5.pdf (accessed 29 December 2019).

IISD (2002). IISD Reporting Services' Complete Coverage of the World Summit on Sustainable Development 26 August–4 September 2002, Johannesburg, South Africa.

Available from: http://enb.iisd.org/2002/wssd/WSSDcompilation.pdf (accessed 29 December 2019).

Independent Group of Scientists appointed by the Secretary-General (2019). Global Sustainable Development Report 2019: The Future is Now – Science for Achieving Sustainable Development. United Nations, New York.

Jamail D. (2017). Interview: Learning to See in the Dark Amid Catastrophe: An Interview With Deep Ecologist Joanna Macy. Available from: https://truthout.org/articles/learning-to-see-in-the-dark-amid-catastrophe-an-interview-with-deep-ecologist-joanna-macy/ (accessed 23 November 2019).

King M. L. (1963). I have a dream. Available from: www.archives.gov/files/press/exhibits/dream-speech.pdf (accessed 28 December 2019).

Kohlrieser G. (2012). Care to Dare. Jossey-Bass, Plano, TX, USA.

Lovelock J. (1979). Gaia, a New Look on Life on Earth. Oxford University Press, Oxford, UK.

Lyons O. (2004). The Ice is Melting. Twenty-fourth annual E. F. Schumacher Lectures. Schumacher Center for New Economics, Hillsdale, MA, USA.

Meadows D. (1993). Down to earth, speech delivered at a conference in Costa Rica. Available from: www.youtube.com/watch?v=bxowxs22jFk (accessed 30 December 2019).

Merriam-Webster. Definition of stewardship. Available from: https://www.merriam-webster.com/dictionary/stewardship (accessed 23 November 2019).

Michel J. B., Shen Y. K., Aiden A. P., Veres A., Gray M. K., Brockman W., The Google Books Team, Pickett J. P., Hoiberg D., Clancy D., Norvig P., Orwant J., Pinker S., Nowak M. A. and Aiden E. L. (2011). Quantitative Analysis of Culture Using Millions of Digitized Books. Science, New York, USA (see also https://books.google.com/ngrams (accessed 29 December 2019)).

Naess A. (2008). Ecology of Wisdom. Counterpoint, Berkeley, CA, USA.

Ostrom E. (2009). Nobel Prize Lecture by Professor Elinor Ostrom receipt of the 2009 Sveriges Riksbank Prize in Economic Sciences. Available from: www.youtube.com/watch?v=T6OgRki5SgM (accessed 29 December 2019).

Rahaman M. M., Varis O. and Kajander T. (2004). EU water framework directive vs. integrated water resources management: the seven mismatches. *Water Resources Development*, **20**(4), 565–575.

Rifkin J. (2009). The Empathic Civilization: The Race to Global Consciousness in a World in Crisis. TarcherPerigee, New York, USA.

Rilke R. M. (1929). Letters to a Young Poet. Insel Verlag, Leipzig, Germany.

Rockström J., Falkenmark M., Lannerstad M. and Karlberg L. (2012). The Planetary water drama: dual task of feeding humanity and curbing climate change. *Geophysical Research Letters*, **39**, doi: 10 1029/2012GL051688.

Saint-Exupéry A. (1943). The Little Prince. HMH Books for Young Readers, Boston, MA, USA.

Scharmer O. (2013). Leading from the Emerging Future: From Ego-System to Eco-System Economies. Berrett-Koehler Publishers, San Francisco, CA, USA.

Shiva V. (2007). The Nine Principles of Water. An essay in Rastello, E. and Sipalla, H. (2007). World Water Crisis – A challenge to Social Justice. St. Pauls Communications/Daughters of St Paul, Boston, MA, USA.

Smith A. (1759). The Theory of Moral Sentiments. Available from: https://static1.squarespace.com/static/56eddde762cd9413e151ac92/t/56fbaba840261dc6fac3ceb6/

1459334065124/Condensed_Wealth_of_Nations_ASI.pdf (accessed 28 December 2019).

Soeltoft P. (2017). Ten things leaders can learn from Kierkegaard/10 ting ledere kan lære af Kierkegaard. Akademisk Forlag, Copenhagen, Denmark.

Solomon S. (2011). Water: The Epic Struggle for Wealth, Power, and Civilization. Harper Perennial. New York, NY, USA.

Strang V. (2004). The Meaning of Water. Bloomsbury, London, UK.

Stubberup M. (2004). Hjertebøn og Østkirkens mystikere (Heartprayer). Borgen, Copenhagen, Denmark.

Thoreau H. D. (1856). From Henry David Thoreau's journal. Available from: https://hdt.typepad.com/henrys_blog/2012/01/january-5-1856.html (accessed 23 November 2019).

UN (1972). Declaration of the United Nations Conference on the Human Environment. Available from: www.ipcc.ch/apps/njlite/srex/njlite_download.php?id=6471 (accessed 23 November 2019).

UN (2015a). Sustainable Development Goals. Available from: https://sustainabledevelopment.un.org/ (accessed 29 December 2019).

UN (2015b). Transforming our world: the 2030 Agenda for Sustainable Development. Available from: https://sustainabledevelopment.un.org/post2015/transformingourworld (accessed 23 November 2019).

Varela F. and Maturana H. (1980). Autopoiesis and Cognition: The Realization of the Living. Springer Verlag, Berlin, Germany.

Water Foot Print (2019). Water footprint. Available from: https://waterfootprint.org/en/water-footprint/ (accessed 23 November 2019).

Water Research Commission (2013). Biomimicry for constructed wetlands. Looking To Nature For Solutions On Water Treatment. Available from: www.wrc.org.za/?mdocs-file=43146 (accessed 28 December 2019).

Weisman A. (2007). The World Without Us. St. Martin's Thomas Dunne Books, New York, USA.

Weisner J. and York H. (1964). National Security and the Nuclear-Test Ban. *Scientific American*, **211**(4), 27.

WFN (2019). Water Footprint Network website. Available from: https://waterfootprint.org/en/ (accessed 29 December 2019).

WHO (2000). Millennium Development Goals. Available from: www.who.int/topics/millennium_development_goals/about/en/ (accessed 29 December 2019).

Wilber K. (1996). A Brief History of Everything. Shambhala, Boulder, CO, USA.

WMO (1992). The Dublin Statement on Water and Sustainable Development. Available from: www.wmo.int/pages/prog/hwrp/documents/english/icwedece.html (accessed 29 December 2019).

Woods M. C. (2010). 'You can't negotiate with a beetle': environmental law for a new ecological age. *Natural Resources Journal*, **50**(1), 167–210.

Woolf V. (1976). Moments of Being: Unpublished Autobiographical Writings. Mariner Books, Boston, MA, USA.

World Water Council (2000). 2nd World Water Forum, The Hague, March 2000 from Vision to Action. Available from: www.worldwatercouncil.org/en/hague-2000 (accessed 2. December 2019).

# Index

IWA Publishing's authorised EU representative for General Product
Safety Regulations is Diane D'Arras, 15 rue Duret, 75116 Paris,
France, e-mail: safety@iwap.co.uk.


Printed and bound by CPI Group (UK) Ltd, Croydon, CR0 4YY

16/06/2026

02140404-0001